T0234548

THE ETHICAL PRIMATE

'A highly interesting and wide-ranging book.'

Explorations in Knowledge

'Commonsense philosophy of the highest order.'

Financial Times

'Her analyses of freedom, fatalism, and reductivism are clear and cogent.'

Stephen Clark, *New Scientist*

In her new book, Mary Midgley argues that the crude isolation of mind and body in reductionist scientific theories still causes painful confusion. Such theories ignore the importance of the higher human faculties and do not leave any room for a realistic notion of the self.

This reductionism is in fact unscientific. There is not just one single legitimate explanation. There are as many answers as there are viewpoints from which questions arise – subjective and objective, practical as well as theoretical.

Human morality arises out of human freedom: we are uniquely free beings in that we are aware of our conflicts of motive. But those conflicts and our capacity to resolve them are part of our natural inheritance. What matters for our freedom is the recognition of our genuine agency, our slight but nevertheless real power to grasp and arbitrate our inner conflicts.

Mary Midgley is a moral philosopher with a special interest in evolution and the relations between science, religion and everyday life. Until her retirement in 1980 she was Senior Lecturer in Philosophy at the University of Newcastle.

OTHER BOOKS BY MARY MIDGLEY

BEAST AND MAN
The Roots of Human Nature

HEART AND MIND
The Varieties of Human Moral Experience

ANIMALS AND WHY THEY MATTER

WOMEN'S CHOICES
Philosophical Problems Facing Feminism
(*with Judith Hughes*)

WICKEDNESS
A Philosophical Essay

EVOLUTION AS A RELIGION

WISDOM, INFORMATION AND WONDER

CAN'T WE MAKE MORAL JUDGEMENTS?

SCIENCE AS SALVATION
A Modern Myth and its Meaning

THE ETHICAL PRIMATE

Humans, Freedom and Morality

Mary Midgley

Routledge
Taylor & Francis Group

LONDON AND NEW YORK

First published 1994
by Routledge
2 Park Square, Milton Park, Abingdon, Oxon, OX14 4RN
605 Third Avenue, New York, NY 10017

*Routledge is an imprint of the Taylor & Francis Group,
an informa business*

Paperback edition 1996

Copyright © 1994 Mary Midgley

Typeset in Galliard by
Florencetype Ltd, Kewstoke, Avon

British Library Cataloguing in Publication Data
A catalogue record for this book is available from the British Library.

Library of Congress Cataloguing in Publication Data
A catalogue record for this book is available from the Library of Congress.

ISBN 13: 978-0-415-13224-4 (pbk)

Publisher's Note
The publisher has gone to great lengths to ensure the quality of this reprint
but points out that some imperfections in the original may be apparent.

CONTENTS

Acknowledgements   ix

Part I The problem

1 INNER DIVISIONS   3

2 MISGUIDED DEBATES   13

Part II The reductive enterprise

3 GUIDING VISIONS   27

4 HOPES OF SIMPLICITY   43

5 CRUSADES, LEGITIMATE AND OTHERWISE   52

6 CONVERGENT EXPLANATIONS AND THEIR USES   63

7 TROUBLES OF THE LINEAR PATTERN   71

8 FATALISM AND PREDICTABILITY   80

Part III The sources and meaning of morals

 9 AGENCY AND ETHICS   95

10 MODERN MYTHS   109

11 THE STRENGTH OF INDIVIDUALISM   121

12 THE RETREAT FROM THE NATURAL WORLD   128

13 HOW FAR DOES SOCIABILITY TAKE US?   136

14 THE USES OF SYMPATHY   141

Part IV What kind of freedom?

15 ON BEING TERRESTRIAL   157

16 WHAT KIND OF BEINGS ARE FREE?   169

17 MINDS RESIST STREAMLINING   177

Notes   185

Index   192

ACKNOWLEDGEMENTS

This book grew out of three earlier articles which still surface in it from time to time, though they have been considerably submerged by rewriting. I would like to thank those who formerly published them for allowing them to reappear. For permission to re-work 'Reductivism, fatalism and sociobiology', which underlies a good deal of Part II, I have to thank the editors and publishers of the *Journal of Applied Philosophy*, where it appeared in vol. 1, no. 1, for 1984. Peter Singer and Messrs Basil Blackwell have kindly allowed me to use material from 'The origins of ethics', which formed a chapter in *The Companion to Ethics* in 1991 – material that has been built into Part III of this book. And I must thank the Cambridge University Press for releasing 'On being terrestrial', an article which appeared in their volume *Objectivity and Cultural Divergence* in 1984 and has here been absorbed into Part IV.

Part I
THE PROBLEM

1

INNER DIVISIONS

MAKING SENSE OF US

HUMAN MORALITY is not a brute anomaly in the world. Our moral freedom is not something biologically bizarre. No denial of the reality of ethics, nothing offensive to its dignity, follows from accepting our evolutionary origin. To the contrary, human moral capacities are just what could be expected to evolve when a highly social creature becomes intelligent enough to become aware of profound conflicts among its motives.

In a way, this may seem obvious. After all, our actual characteristics have got to make evolutionary sense, and though there are still wide disputes about the details of evolutionary theory, all serious scientists agree on the central fact of our descent from other social animals. Yet clearly many people today are still most uneasy about it. The vast tide of print that has flowed for more than a century round the shores marked 'Darwinism' has never really succeeded in making this particular sandbank navigable.

For instance, my title may well disturb some readers. Is it perhaps unduly reductive to call *Homo sapiens* an ethical primate? It may seem so, in the same sort of way in which the title of Desmond Morris's book *The Naked Ape* did twenty-five years ago. Critics complained then, with some reason, that nakedness was not a specially significant distinguishing property of our species, and also that humans were not literally apes. What I am saying now, however, is literal fact. Most of us agree that we literally *are* primates who possess ethics, and also that the capacity for ethics is an extremely significant property of our species.

Neither of these propositions seems really implausible, yet – as I am suggesting – current ways of thinking still make it hard to bring them together. This book is one more instalment in an attempt to make that process a little easier. That is needed, not just so that we can accept our history but also (what matters more) so as to help us towards a more integrated notion of ourselves. In trying to map the neglected connection between the two brightly-lit patches that we normally look at on our vast, confusing, half-lit intellectual landscape – the scientific world-view and the world as we see it every day – I am also trying to bridge a similar gap that splits our idea of our own nature.

I began that reconciliatory attempt in my first book *Beast and Man*[1] and I have continued it in *Heart and Mind*[2] and in *Wickedness*[3] – indeed to some extent in all my books. This present book concentrates on problems about the nature of inner freedom and its relation to personal identity. These – along with the moral issues discussed in *Wickedness* – seem to me the most central business left over from *Beast and Man*. Here, as elsewhere, I look primarily at the howling mistakes in which we are now involved, and try to suggest ways of curing them. Attention to the imaginative aspect of those mistakes – to the symbolism by which they pervade our lives – seems needed as well as attacks on the official arguments involved. I want to examine the way in which over-simple ideas about the relation between the inner and the outer standpoint, between subjective and objective, make it seem that we cannot have certain crucial kinds of freedom. I want to see how we can avoid these misleading kinds of reduction without becoming unintelligible.

In this project, it has seemed necessary to revisit some points discussed in *Beast and Man*. I have found that these ideas still evidently startle many people, and may yet need some time to become familiar. But I hope that readers who are at ease with them can easily skip these passages. Apart from that, some of the things I am saying are certainly obvious. But then, retrieving obvious things that have got lost can be important philosophical business, and I think it is so here. This whole area is so difficult and so easily generates paradox that even very obvious truths often get forgotten in it.

DIVERGENT VISIONS

Darwin himself worried deeply about the evolutionary meaning of morality, and many other people have shared his concern. They have tried hard to find some intelligible relation between human moral consciousness and the patterns underlying the long development of life. But the people doing this quickly divided themselves into two extreme camps – one reductive and the other obscurantist. That split, which persists strongly to this day, is not just a debate between theorists. It works at an everyday level and obscures matters vitally important to our lives. This book will try to attend to this feud in some detail, to light up the ways in which it misleads us, and to suggest how we can deal with them.

As with many such feuds, the two sides are kept apart not just by their views, but more deeply by a determined contrast of style and tone. The reductive party – in its early stages embodied in Social Darwinism – likes to shock. It is positively pleased to sound harsh, strident and paradoxical, since it views these qualities as marks of realism. Its basic project is to unite humans to the rest of the biosphere by following Procrustes, by paring all human peculiarities down to a size which fits easily into the supposedly universal evolutionary pattern.

Even before Darwin, that pattern was seen as one of cut-throat competition for survival, a model set by an older reductive approach in political philosophy. Nineteenth-century Social Darwinists were following Hobbes's lead in preaching that human conduct was wholly directed by self-interest. But this kind of crude psychological egoism has never been wholly convincing, in spite of its political uses. Accordingly, in the last few decades the sociobiologists have moved the thesis away from empirical falsification to the safer, more metaphorical doctrine of gene-selfishness. They have not, however, softened its crude, cocky, omniscient, debunking tone at all. Thus E.O. Wilson:

> Human behaviour – like the deepest capacities for emotional response which drive and guide it – is the circuitous technique by which human genetic material has been and will be kept intact. Morality has no other demonstrable ultimate function.[4]

> The organism is only DNA's way of making more DNA.[5]

Similarly Richard Dawkins:

> We are survival machines – robot vehicles blindly programmed to preserve the selfish molecules known as genes. This is a truth which still fills me with astonishment.

> We, and all other animals, are machines created by our genes
> . . . Like successful Chicago gangsters, our genes have survived,

in some cases for millions of years, in a highly competitive world. I shall argue that a predominant quality to be expected in a successful gene is ruthless selfishness ... We are born selfish.[6]

The opposing party – the party that stresses mystery and discontinuity with other species – has been formed in direct response to this attack, and it has therefore tended to speak in the style that always answers such challenges. Its tone is grieved, parental, mature, concerned, in fact genuinely outraged. It has replied that human morality is so totally unlike the travesty of it offered by the reducers as to be evidently disconnected from everything else in evolution. Morality can therefore never be brought into any kind of intelligible relation with its earthly surroundings. It is a distinct, unassimilable pattern at odds with all else on this planet, and perhaps with everything in the universe.

THE WRONG DRAMA

This dispute is often seen as a simple tribal clash between scientists and Christians. But that pattern is doubly mistaken. On one side, Social Darwinism and its descendants have had very little connection with physical science. They were born in economics and nurtured by political theory. They have always owed much of their force to thoughts about commercial freedom – a connection which is still strongly marked, now that they have fed back into biology, by the sociobiological language of 'investment'.

On the other side, the resistance to seeing morality as merely a weapon in an egoistic contest for survival is quite independent of religion. Religious people have certainly raised these objections, but they have not been alone in doing it, nor have they needed special religious grounds for their protest. Strongly pro-Darwinian and anti-religious sages have stressed exactly the same point. In fact, its most vigorous early spokesman was 'Darwin's bull-dog', T.H. Huxley. Huxley reacted explosively against Herbert Spencer's complacent evolutionary ethic, which taught not only that evolution was for the best, but that it was the sole guide to morals, since what was right was simply whatever furthered evolution through 'the survival of the fittest', giving this as a reason for not helping the unfit poor.[7] Much though Huxley hated Christianity, he hated this reduction of morality to evolutionary processes far more. So he declared uncompromisingly that:

Ethical nature, while born of cosmic nature, is necessarily at enmity with its parent ... The ethical process is in opposition to the principle of the cosmic process, and tends to the suppression of the qualities best fitted for success in that struggle ... [Man must therefore be] perpetually on guard against the cosmic forces, whose ends are not his ends ... Laws and moral precepts are directed to the end of *curbing the cosmic process*.[8] (Emphasis mine).

Darwin's collaborator Alfred Wallace objected quite as strongly, and, not being anti-religious, he proposed a supernatural solution for this emergency, saying that God had created human mental powers separately in the course of an otherwise natural evolution. But Huxley's original difficulty arises quite independently of any such efforts to resolve it. Important though the religious angle may be for other reasons, it is not the source of the trouble here. That source lies deeper. There is a real, insuperable difficulty in making the actual facts of human life fit into crudely reductive pictures such as the Social Darwinist one.

SECULAR SEPARATISM
AND THE MINIMAL SELF

Throughout this book I shall concentrate on the problems that arise within a secular approach rather than on those raised within the religions. This is not because questions involving religion are unimportant, but because, once they are raised, people tend to hare off after them as if merely getting rid of religion would solve all our problems. It will not. Secular thinkers as well as religious ones need to find an alternative to unrealistic reductive views of personal identity. They have often felt that they could only meet this difficulty by separating the essential self altogether from the biological body. That body is, after all, certainly a part of the cosmic process which alarmed Huxley so much. The formula for a drastic separation had already been shaped by Descartes, who cleared the conscious mind out of the physical sciences by ruling that it was a non-physical substance, a mere passenger in the body.

This drastic division has indeed often been thought of as part of Christian thought, though it actually conflicts with much of the Christian tradition, notably with the doctrine of the resurrection of the body. That doctrine means that soul and body, at a deep level, are one. As Arthur Peacocke puts it, both in the Old and the New

Testament 'a human being is regarded as a psycho-somatic unity, a personality whose outward expression is his body and whose centre is his "heart", "mind" and "spirit" '.[9] What is envisaged at the resurrection is not, then, that corpses revive, but that the soul develops other faculties to replace the body. Though this idea about mind–body relations may be puzzling as regards the next life, for the present life it seems to be a good deal more intelligible than the kind of neo-Cartesian separatism that has prevailed in recent secular thinking.

We will look at this separatism more closely later, but a few examples may be useful now. One case of it is Jean-Paul Sartre's Existentialist morality. That morality treats the human will as a spontaneous, independent force, completely detached from all natural motives and capable of opposing them all. (This idea of its independence seems also to have been implied, though in more muted terms, in the accounts given by some emotivist and prescriptivist philosophers). A second case is the prolonged refusal of many social scientists to admit that genetic factors can have any influence at all on people's mental lives, especially in such crucial matters as sex roles. About sex differences, the orthodox doctrine for several decades was that, as Dr John Money put it in 1955,

> Sexuality is undifferentiated at birth and ... it becomes differentiated as masculine or feminine in the various experiences of growing up.[10]

Only very gradually is this dogmatic insistence now changing. It has of course had the honourable political motive of supporting women's equality. But equality is not sameness. A belief in sameness here is both irrelevant to the struggle for equal rights and inconsistent with the facts. It ignores massive evidence of sex differences in brain and nerve structure occurring long before birth, and also of behavioural differences which are evidently independent of culture and sometimes contrary to it. It amounts to an extraordinarily abstract notion – evidently held on moral grounds – of the original human being as something neutral, sexless and indeterminate, something wholly detached from the brain and nervous system.[11]

Nicholas Dent, discussing most helpfully this strange contraction of the self and its supposed connection with the idea of freedom, cites an advertisement for the 'fifth and final volume' in the 'award-winning series *A History of Private Life*' in which (it is promised) 'nine noted historians chart the remarkable inner history of our times . . . when personal identity was released from its moorings in gender, family, social class, religion, politics and nationality.'[12] Dent comments,

> One cannot represent all specific circumstances which impinge on an individual, all attributes ascribed to an individual . . . as limitations on, constraints upon, the self. For what then remains, to comprise the concrete actuality of the self's existence, is nothing, or almost nothing, a will without grounds, a power of choice without objectives.[13]

Hegel (he remarks) called this 'the freedom of the void', and it is surely not what anybody really aims at. But the idea of it is easily reached as we go on counting up the various external influences that we might sometimes want to disown. As Dent says,

> One is apt, particularly under the pressure of more and more discoveries about formative influences, inherited and circumstantial, on intelligence, personality, temperament etc., to represent more and more as external to the constitution of oneself, as mere barnacles that encrust the surface of the soul.[14]

This process goes on today on an industrial scale, but curiously little attention is paid to the shrinking image of personal identity that it produces.

VIRTUAL PEOPLE

A third, still more striking example of separatism is the position that comes so naturally to many champions of artificial intelligence – the belief that not only will computers (or their programmes) one day be made conscious in exactly the sense in which human beings are conscious, but that human beings are already, in some fairly literal sense, themselves programmes run on computers made of meat. Many people today find this kind of proposition so obvious that they do not even bother to argue for it, but confidently put the burden of proof on anyone who suggests otherwise. It seems worth while to illustrate this strange position by a longish quotation from a recent book by two very distinguished cosmologists, John Barrow and Frank Tipler:

> An intelligent being – or, more generally, any living creature – is fundamentally a type of computer . . . The really important

9

part of a computer is not the particular hardware, but the program; we may even say that a human being *is* a program designed to run on particular hardware called a human body ... *The essence of a human being is not the body but the program which controls the body,* we might even identify the program which controls the body with the religious notion of a *soul,* for both are defined to be non-material entities which are the essence of a human personality. In fact, defining the soul to be a type of program has much in common with Aristotle [*sic*] and Aquinas' definition of the soul as 'the form of activity of the body' ... When atoms disappear human bodies will disappear, but programs capable of passing the Turing test need not disappear. *An intelligent program can in principle be run on many types of hardware,* and, even in the far future of a flat Friedman universe, matter in the form of electrons, positrons and radiation will continue to exist.[15] (Longer emphases mine)

These authors are certainly lucky in not needing to answer any comments from Aquinas, nor from that most biological of philosophers, Aristotle. Their project is to make the human race immortal. They think this can be done by transferring human minds to computers so as to prolong their existence into an epoch when there will be no other organized matter at all – a time when there will be nothing to do except (presumably) to consider and communicate abstractions. In the absence of anything to talk about, it is not even clear what the topics of these conversations will be, except perhaps mathematics.

This particular enterprise is certainly in some ways an unusual one. But the assumptions that make it look possible are not personal quirks of these authors. They are widespread throughout the artificial intelligentsia. Science-fiction, which has always been naïvely Cartesian, has acclimatized its readers to this bizarre way of thinking. For our present purpose, what matters is not so much the wild initial assumption that consciousness could be transferred to such machines. It is the further assumption about values, the assumption that the life which they would then live – a life without sense-perception or emotion or the power to act, a life consisting solely in the arrangement of abstract 'information' – would be a human life, or indeed anything that could intelligibly be called life at all.

The extra fact that in these particular circumstances there would be nothing left to talk *about* certainly lights up the oddity of the picture. But it does not create that oddity. Even if conscious computers are imagined as existing in an ongoing world, they still have to be credited with a kind of consciousness that operates in this extraordinary vacuum, detached from perception, feeling and action.

I do not think the notion of consciousness makes any sense in these circumstances. Everything with which consciousness normally occupies itself has been removed. It is important to distinguish the idea of this vacuous state from the idea of the sort of heightened consciousness that mystics have aspired to. For the mystics – to speak roughly of what requires subtlety – ordinary experience is transcended through spiritual efforts which make possible higher and more complex kinds of perception, emotion and action. But this is seen as an *advanced* form of experience, one which can only arise after much hard work has been done at the ordinary human level. It is not recommended as a technical fix, a cheap substitute supplied by the engineers which can enable us to dispense with that level.

IS THIS SCIENCE?

These are, of course, three very varied examples of mind–body separatism. We will come back to the topic later. At present I want to stress just two things about them. The first is how naturally they arise today. Their very variety testifies to the wide spread of this isolation of the mind. It shows how extraordinarily easy it is for many sorts of people now to think of their essential selves as something discontinuous with the body which roots them in the evolutionary process.

Second, it is interesting to see how the language in which this separatism is expressed is changing. Sartre, writing in 1945, saw his views as something quite detached from natural science, indeed as a protest against using scientific concepts at all to talk about the essential self. Clearly much influenced by Descartes, he wanted to exempt the autonomous will from the domain of science as fully as Descartes had exempted the soul. By contrast, modern social scientists are not free to set up this violently dualistic metaphysic. Although they may sometimes defend autonomy on Sartrian lines, they are increasingly driven by the spirit of the age – and indeed by the flow of research money – to speak more or less in scientific language, and to aim at something called the scientific method.

Propounders of artificial intelligence, however, have no inhibitions at all about claiming scientific status. Though 'computer science' is not itself actually a natural science but rather a branch of applied logic and mathematics, its practitioners tend to identify themselves unhesitatingly with the cause of 'science' when battles are being fought, and are in general readily accepted there as allies. Thus Barrow and Tipler observe with satisfaction that their own discussion is now 'based entirely on the laws of physics and computer theory',[16] not, apparently, being aware that much of it is simply bad metaphysics, nor that anyone who is going to lay down the law about the relation of mind and body ought at least to be doing some biology as well.

Thus, modern, secular doctrines separating mind from body have been gradually drawn by the prestige of science into using its rhetoric and some of its concepts, even though more traditional forms of separatism based in the humanities still flourish as well. This does not, of course, necessarily force separatists to adopt the contemptuous tone that we have noted as characteristic of reducers. Barrow and Tipler, to their credit, usually try to avoid that tone and are anxious to bridge the gap between the 'two cultures'. But opinions such as theirs almost unavoidably give occasion for that tone.

The point is not just that calling minds computer programmes can as easily be made to sound shocking as calling them arrangements of tissue or vehicles for genes. The point is the territorial claim that this whole topic belongs not to ordinary people but to experts – that is, scientists, that its place is with questions about protons and the carbon cycle, not with questions about the soul or family life or political freedom. The reductive style, which can now be used on both sides of the controversy, is a style that treats this whole topic as one for people with doctorates, one that dismisses the concepts by which we normally live as mere 'folk-psychology'. This seems to be one of the most serious mistakes that the learned have ever made.

2

MISGUIDED DEBATES

THE TERMS OF THE DEBATE

I HAVE begun by concentrating on the choice of *style* – the drama that is felt to be enacted and the roles people choose in it – because these are not just superficial irrelevancies. They have played a huge part in sustaining this dispute. If we make the great and unusual effort of trying to detach ourselves from that drama and to look at the question more calmly, certain points surely emerge which should be noticed by both parties.

In the first place, the grieved, anti-reductive party needs reminding that unintelligibility is not in itself a gain. Certainly we have no guarantee that the universe, or our own place in it, makes any sense at all. But we can only think about it by assuming that at some level it does do this. Of course particular explanations that fail must be dropped. But that is never a reason for claiming that the gap they leave cannot be bridged. It may well be true, as Thomas Nagel has argued,[1] that the gap between our inner and outer lives – between the subjective and the objective point of view – is in some sense irreducible, and that this gap imposes sharp limits on our chances of unifying our experience. We will come back to this important angle later. But, as Nagel points out, theorists have repeatedly misdiagnosed this particular gap by proclaiming other, more exaggerated kinds of dualism surrounding it – other gaps over which mystery and even warfare must arise – without real justification.

In particular, modern thinking about morality has been prone to surround it with discontinuities that make it look incomprehensible. Since the seventeenth century, Western philosophy has tended to polarize, not only mind and matter but also reason and feeling – to treat these as separate aspects of life, not intelligibly related. Theorists discussing morality accordingly formed the habit of asking simply which of these departments to put it into. Thus, even David

13

Hume, who did want to take feeling seriously, began his *Enquiry* on this subject by asking 'concerning the general principles of morals, *whether they be derived from Reason or from Sentiment*'[2] (emphasis mine) . He then used this forced, artificial choice as part of a wider territorial battle between rationalists and empiricists about the general nature of thought.

This split has a far worse effect when it is applied to the practical thinking involved in morals than it has over factual knowledge, because morality so clearly does involve feeling. The polarizing tradition – which dominated moral philosophy from Hume's time till quite lately – found its most recent expression in the notion that facts and values were conceptually unrelated. This naturally meant that facts about evolution were, along with all other facts, irrelevant to the project of understanding morality. But once we accept our evolutionary history as a general background, it is quite natural and proper to use it in explaining many elements of human life. If we shut morality off from that explanatory pattern of thought, we tend to make its relation to the rest of human life unintelligible, which cannot be an advantage.

WAYS OF SPLITTING THE SELF

That is not just a conjecture. Experience has already shown that, during the twentieth century, the quarantining of morality from topics that obviously relate to it has inclined people to think of it as something vague, irrational, privatized, inarticulate and subjective – in fact, as something beautiful but trivial. For many highly educated people, in fact, ethics is enclosed today in a ghetto that shuts it off altogether from the rest of the intellectual scene. It is liable to share that ghetto with a number of other topics which are hard to fit into currently popular conceptual schemes – awkward items such as consciousness, emotion, creativity and free will.

All of these are readily designated as 'problems'. Just as gravity made a problem for Newtonian mechanics, and the growth of organisms made one for Plato's changeless world, so these phenomena obstruct the tidy reductive schemes that have promised so much for twentieth-century thought. The ghetto exists to make those reductive promises look fulfillable by keeping the awkward items out of the way for the time. Ghetto-users either deny the existence of these topics, translate them into more digestible terms, or – as happens with the fact–value gap – simply shut them off from the rest of thought altogether. Like Procrustes, they assume that the trouble lies in the awkward item. They do not want to ask instead whether there might not be something wrong with the conceptual scheme into which the item refuses to fit.

The price of this refusal is not only a lack of certain useful explanations. It is a split in our notion of ourselves. Ever since the Renaissance the polarizing policy has not just involved enquiries about knowledge, but has also concerned personal identity. It has posed a challenge to each of us to identify our essential selves *either* with reason or with feeling. If the relation between these two is seen as unintelligible, then there is a blank gap within each of us, a chasm across which negotiations can only take the form of conflict.[3]

This chasm has been made particularly hard to cross by being used to dramatize clashes of value. Starting with Plato, many moralists have taken sides with reason, by which they have meant, not just the power of thought, but certain particular motives, such as the wish for order, which seemed to set it going. Conversely, its opponent was seen, not just as feeling in general but as certain particular feelings, often (as with the Stoics) including personal affection, which were to be suppressed.

This confusion, however, grows far worse when that conflict is in some way lined up also with the gap between mind and body. Theorists have always been quick to see that the feelings have a strong bodily basis. They have often been surprisingly slower to notice that the powers of thought have one too, that the intellect must take its shape from the brain and nervous system, and that the motivation to use it does not come out of the air. They have seen thought as somehow standing outside nature. This has made it possible to identify the essential self with reason and to call on it to fight feelings that belong only to that alien beast, the body.

DUALIST QUANDARIES

Thus – to repeat – we are not dealing here just with a question about the past, about the way in which human evolution took place. We are not just asking whether the emergence of our species was gradual or sudden. Some people seem to think that the suddenness of 'punctuated evolution' can isolate us from our past – that a 'hopeful monster' may have appeared which bore all the distinctive marks of modern humanity and no inconvenient heritage from previous ancestors.

But this notion does not help us at all in dealing with the divided self that we now have. And a glance at our present state shows the emptiness of the vision. Nobody really denies that we still have something called our 'animal nature'. Nobody who has watched the parental behaviour of other primates or has seen a human being in the grip of strong passion can doubt that our emotions share a strong common heritage with many mammals. That kinship was recognized long before anybody heard of the theory of evolution, and it is no less obvious today. The question simply is, how are we to understand it and live with it?

Huxley's strategy for handling his war between moral man and immoral nature was to separate the combatants completely. He usually placed the gap – the hiatus dividing the ethical from the cosmic process – between the human mind and body, where it notoriously makes endless difficulties. At other times he set it between civilized and uncivilized humans, which is even worse. But wherever it is placed, this dualism, whether moral or metaphysical or both, blocks thought.

Huxley himself said explicitly that it called for a pre-Socratic metaphysic. He recommended the world-picture of Heraclitus, centring on a primal conflict; 'war is the father of all things'. He represented the world in Manichaean style as sharply divided between good and evil forces. This picture has something in common with the dualism of Eros and Thanatos proposed, for related reasons, by Freud. Both dramas do, of course, make interesting moral and psychological points. But the price they impose for serious acceptance is extortionate. It is not really possible to use a naïve pre-Socratic metaphysic in a sophisticated context. We have no business to accept this kind of obscurantism unless we are really forced to.

The Huxleyan party needs to notice, too, that this claim about a drama of brute opposition is neither more empirical nor less overconfident than the one it was meant to displace. Both equally are just optional imaginative pictures. They are not arguments.

REDUCTION AND REALISM

As for the reductive, debunking party, its trouble is one which very often does infest reduction – it lives too far from the facts. It is no use reducing A to B if you have to misrepresent A so much in doing it that your conclusions plainly don't apply to the world. Rhetorical exaggerations simply are not convincing. Officially, the logical point of reduction is to unify the conceptual scene. But this cannot work unless the details of that scene are still recognizable. Reduction does, of course, often have other secondary purposes such as exposing humbug, annoying the Establishment and cheering up the reducer. These purposes may still be achieved even by ideas which have stopped being realistic. But such aims need to be kept separate from the business of explanation.

In our present case, the reducers' distorted pictures of the human moral scene – especially the crude egoism that usually forms part of them – have been a main reason why many unprejudiced people simply cannot swallow the fact of human evolution at all. These pictures are *not* realistic. They only look convincing while they are being compared with certain special forms of hypocrisy which they were designed to correct – notably, with exaggerated accounts of human virtue. Of course it is true that human beings are not full-time altruists. But since common experience shows that they are not full-time Hobbesian egoists either, we are naturally sceptical when we are told that science says they must be.

At a simple level, this incredulity has been an important source of the current revival of creationism. But it does not only work at that simple level. It also perplexes and confuses much more sophisticated people who do not necessarily invoke religion and have no doubts about the actual history. 'Darwinism' is often seen – and indeed is often presented – not as a wide-ranging set of useful suggestions about our mysterious history, but as a slick, reductive ideology, requiring a kind of philistine obtuseness about all the subtleties of the inner life – requiring us, in fact, to dismiss as illusions matters which our experience shows to be real and serious.

I want, in this book, to suggest that we can find much better ways of understanding this difficult topic. What is chiefly needed is a less abstract, less quarrelsome, more realistic notion of how human freedom actually works. For that we must recognize more fully how complex our motives are, and especially what kind of inner conflicts this complexity involves, both in our own and in other species. The simplified stereotypes used by both the feuding parties just mentioned are the main obstacles to making terms with our history, and thereby with our bodies.

This campaign of over-simplification has, of course, not just been a chance matter. The faith that all complex facts can ultimately be explained by a simple basic structure underlying the physical world has been strong in Western thought from the seventeenth century to our time, and has been a great source of emotional security. It has often been seen as something necessary for rationality itself. It is a habit which has shaped the thought of our age profoundly, and it has certainly held the initiative in this particular debate. It therefore needs full discussion, which will occupy Part II of this book.

ON WORKING IN
THE MARKET-PLACE

The kind of ideas involved here are not the private preserve of academics. Reductionism itself is not just a formal, logical practice. It has always been an ideology as well, an imaginative habit linked with a wide variety of faiths and moral attitudes. It is a temper that has deeply affected the life and thought of innumerable people who are not scholars at all, and who may know very little of science. In examining it, therefore, we need to attend to its full range of imaginative expression.

Philosophers have tended lately to feel that they positively ought not to do this. Though the rise of 'applied philosophy' has lately released them somewhat, they still often feel that professionalism calls on them to deal only with the most abstract forms of thought, leaving its vulgar imaginative, practical and motivational aspects to sociologists or historians of ideas. This cannot be right. The huge and turbulent area of everyday thinking – the thinking by which we live – does not just need to be described and recorded, it needs also to be criticized, to be examined, to be taken seriously, even in its most lurid or odious manifestations. It needs to be noticed as thought and not just as noise.

Like many other barriers now growing up round specializations, the philosophical purism that neglects this work is surely doomed because what it keeps is deeply dependent on what it neglects. No thought, however abstract and purified, is an island. No iceberg exists only above water. The more abstract forms of argument emerge organically out of the imaginative background that they express.

Earlier philosophers in the great European traditions usually understood this. They did not attempt this kind of disinfection. They dealt with the dreams and visions that guided the thought of their various epochs. Plato set an excellent example here in his use of myth – especially in his discussion of how myths connect with metaphysic in the *Theaetetus*. Until lately, most of his successors followed this example in their various ways. I do not think any good philosophic reason has ever been brought forward for the recent retreat into specialization, though G.E. Moore and Bertrand Russell did give some bad reasons for it. On the whole, this retreat has resulted rather from drifting with the tide of changing academic practice – from following the increase of specialization elsewhere – than from any clear decision.[4]

Inside every real philosopher there is not only a lawyer but also that lawyer's client, somebody with a substantial idea to express. Unfortunately, schools of philosophy do not necessarily keep these two on speaking terms and tuition very easily tends to concentrate on training the lawyer without listening to the client, a policy which can end by producing a lawyer who does not know he needs a cause. European philosophy, which was born under the shadow of the Athenian lawcourts, has always carried their mark, but never so deeply as it does at present.

Current philosophers still claim their inheritance. They have not yet changed the brass plate on their door to read 'consultant linguistic analyst' or 'consistency inspector'. Until they do that, they should surely use all the resources that this kind of work needs. That involves taking popular works extremely seriously and examining disreputable arguments quite as carefully as respectable ones, something which I shall do throughout this book.

It also means, unfortunately, continuing to take seriously prophets whose views ought long ago to have died with them. Nearly all of us, even those who on paper are most scrupulously post-post-modern, still conduct a good deal of our life and thought by obsolete ideas. Old hats may be revamped but they remain in use. Seductive errors are Hydra-headed monsters that need killing a thousand times. Some of these resilient monsters, such as those produced by Herbert Spencer and B.F. Skinner, can at times produce despair. Can it be any good to keep on battering at ideologies that have survived so many conclusive refutations?

I think that we have to keep on trying, and that perseverance does shift them in the end. I have tried to attack them from a slightly distinctive angle. I concentrate, not so much on refuting particular arguments as on pointing out the wider imaginative landscapes that have made them look plausible, the visions that shape the thought behind them. I want to map the whole terrain in a way that can suggest a way out of various dead ends in which people easily get trapped.

*   *   *

THE SHAPE OF THE BOOK

It seems best to start this discussion by considering the general meaning of reductivism. Accordingly, the second part of this book notes the colourful ideologies that have gathered round the reductive programme. It traces how they have narrowed our powers of explaining human conduct, and in particular how the reducers' refusal to recognize that most crucial piece of material, the first-person view of agency, has blocked our chances of understanding human freedom. This Part ends constructively by sketching, in Chapter 6, an alternative, non-reductive pattern of explanation, one that is pluralistic but rational and usable; indeed familiar.

The third Part asks how, once we have this more flexible explanatory pattern, we can bring the objective and subjective points of view together in a way that makes the rise of morality intelligible. It notes the difficulties that certain influential modern myths, particularly egoist patterns related to the Social Contract, have raised for this enterprise. As an antidote for these myths, it examines Darwin's more realistic derivation of morality as a response to natural conflicts of motive. This is a unique response, made both necessary and possible by humanity's uniquely clear awareness of those conflicts. Yet it is still one continuous with the responses of other animals, because the conflicts themselves are so. Inner conflict itself, of a kind that is to be expected in an evolved creature, is thus seen to be central to freedom and to the morality by which we try to manage it.

Part IV looks more closely at the notion of human freedom itself, asking what it involves. Here we notice the strange process by which the part of the personality that is deemed to be free has, in modern times, been gradually shrinking. Freedom has been delegated to the will, which has itself become so abstract in writers such as Nietzsche and Sartre as almost to constitute an alien ruler over the rest of the character. The essential self thus preserves its dignity by leaving the earth altogether, by disowning all feelings that may seem to be rooted in the body. Though the religious context has changed, Descartes' radical division between mind and body still retains its full dramatic force.

That separation of mind from body, which surely plays a great part in the persisting discomfort over evolution, is not confined to these particular moralists. It is found in many other contexts today, notably in artificial intelligence and in the social sciences. In all these areas, human dignity is chiefly seen as demanding freedom *from* the body – a notion which is not just grotesquely over-inflated but quite misdirected. The way in which I want to avoid this misdirection may be unfamiliar, so it may be best to sketch it out briefly here.

ON BEING FREE AS A WHOLE

Humans do not enter their body as a separate component, an alien driver or pilot who will steer it around. To the contrary, what really is distinctive about human freedom is an individual's ability to act, in spite of many inner divisions, as a whole. This ability is certainly surprising. In recent times, theorists have largely busied themselves in pointing out the ways in which each of us is *not* a self-contained whole.

On the one hand, they have stressed our continuity with the world around us – both socially, through the influence of our cultures, and physically, by the engulfment of our bodies in the rest of nature. On the other, they have remarked on our inner divisions. Freud and other psychologists have stressed how little we understand our own motives and what deep conflicts perturb even those motives of which we are more or less conscious. More recently, neurologists have noted how fragmentary and often conflicting are our perceptions, those elements of experience that we thoughtlessly treat as a solid, given, reasonably harmonious whole. Indeed, the philosopher Daniel Dennett has lately written a large and enthralling tome about this fragmentation,[5] apparently concluding from it that each of us is in some way mistaken in supposing him- or herself to be conscious as a single individual at all.

What is happening here? Clearly, these theorists are quite right to oppose over-simple, over-confident notions of human individuality – notions which have indeed had a lot of influence. They are right to point out that humans have not got the simple, straightforward wholeness of billiard balls. That was why David Hume declared that the self was no single thing. Instead, he said, each person was just 'a bundle or collection of different perceptions which succeed each other with inconceivable rapidity, and are in a perpetual flux or movement'. 'For my part', Hume added,

> when I enter most intimately into what I call *myself*, I always stumble on some particular perception or other, of heat or cold, light or shade, love or hatred, pain or pleasure. I never can catch myself at any time without a perception, and never can observe anything but the perception.[6]

Hume's readers have naturally asked him what is the string around the bundle – a typical misleading mechanistic metaphor – and also *who* it is that thus enters, stumbles and fails to catch or observe? No doubt answers of some kind can be found to these queries. But the trouble goes much deeper. Hume treats his question about unity as though it were a factual question with a single answer, and it is not. His whole approach – which is still often followed today – is mistaken. *Whether something is a whole always depends on the point of view from which you look at it.* This does not make the issue just a matter of taste. It simply means that that point of view has to be specified.

We cannot see the wholeness of a dance or a song by looking at its parts, nor the wholeness of a leaf by looking at it down a microscope. Some dances, songs and leaves are more unified, more complete than others, and unified in different ways. But observers who do not understand the point of these kinds of dances, songs or leaves cannot see this difference even if they get to an appropriate distance. Such observers may well say 'nothing here but a lot of stamping around or a series of yells or a mess of spikes.' And again, observers who do understand these things may have different opinions about this issue according to their different views on what is the central point of a song, dance or leaf.

This does not at all mean that wholeness in these cases is illusory. It means that *all judgments of wholeness depend on value-judgments* – on distinctive ideas of what the kind of whole in question is meant to be or do. Once we have a clear idea about that, neither the compositeness of the thing in question nor its dependence on the surrounding scenery need compromise its wholeness.

What kind of a whole, then, might human beings be? First, they are organisms. And all organisms have their own kind of individuality. Though they all depend deeply on what is around them, each of them is discontinuous from it in so far as it has its own *conatus*, its tendency to preserve and perpetuate its own kind of being in a way that stones or stars do not. Leaves have their own wholeness as well as trees, but each kind needs to be defined with reference to the wider context.

Second, among organisms, human beings are animals. This means that they can act positively and deliberately in a way that plants do not. The point at which this kind of distinctiveness sets in may be uncertain, and people who are sceptical about animal consciousness sometimes postpone it to an oddly late stage. But nobody now doubts that it occurs before we reach *Homo sapiens*.

The next step is also uncertain in its place of onset, but not in its meaning. Third, humans are social animals. Not only do they act but they are aware of the actions of others *as actions* – as significant performances. They receive those actions not just as events but personally, as threats or alarm-calls or invitations, insults or gestures of conciliation or expressions of love from another distinct individual. They respond to these acts, not just by immediate answers, but by building up lasting and complex relations with that individual. And these relations are far more important in their lives than the particular acts. If they lose that individual, their life may lose its meaning. They may pine and sometimes die. In order for all this to happen, it is of course necessary that they themselves should also act as distinct individuals, and should feel themselves to be so. That feeling evidently is the ground-bass of experience for all social creatures, manifesting itself in personal interaction throughout their lives right from their earliest responses as infants.

AND HUMAN UNIQUENESS?

What, however, about the fourth stage – the one which has so often been stressed on its own and loaded with far more work than it can possibly handle? What distinguishes humans among other social animals? Well, this distinctive point is indeed complex, and ought not to be simplified as it constantly has been by people anxious to exalt some particular human trait.[7] But for our present question – for its relevance to personal identity – that point is not too hard to state and is certainly impressive enough.

Human beings are distinctive in being enormously more aware than other creatures both of their individuality and of the factors, both inside and outside them, that compromise it. They can think and talk and argue about these things, so they can share much of their experience and help each other with these problems. They can be aware of forces that are prolonging or changing their ways of life and they can, if they wish, direct their efforts to supporting or resisting them.

Our unity as individuals is not something given. It is a continuing, lifelong project, an effort constantly undertaken in the face of endless disintegrating forces. We, as well as every mouse and every apple-tree, struggle for this wholeness as best we can throughout our lives, undiscouraged by endless obstacles. And we struggle in quite a different style from them because our struggle is conscious.

Who, then, is it within us that so struggles? Pronouns cannot really deal with this kind of question, and they have made constant trouble whenever this or that inner aspect of us has had to be personified. But the answer I want to stress is the obvious and neglected one – It is the whole. The constitution of the creature demands this struggle. Integration is not the responsibility of a disapproving colonial governor, a rational super-ego, nor of a perverse psychological observer determinedly stumbling in the wrong direction. It flows from the shape of our active nature itself. It takes place, not against the laws of our natural being, but according to them. I think this is what Locke meant when he insisted that it is the man, not the will that is free.[8] Much of his discussion seems to me sensible, though he got into trouble by trying to keep using the weasel word *determine*. Locke's account has usually been treated as evasive, because it makes little of certain difficulties which people have often thought overwhelming. But as I shall explain, I think that Locke was often right there.

What about other animals? We know little of their inner lives and ought not to dogmatize about it. But, crudely speaking, though they do share our struggle to harmonize conflicting motives, they plainly do not have anything like our power of dealing with it by standing back from their various motives, by taking the point of view of the whole, and trying to make some kind of balanced decision. They have other distinctive powers, but not this one. That difference is indeed sufficiently striking to make human life radically different and to furnish us with such unique dignity as we actually have. But the incompleteness of this control, along with the fact that both our motives and our minds are still gifts of nature rooted in the body, constitutes a deep continuity with our relatives. This continuity can be very helpful to our understanding of ourselves if we can manage to use it without being ashamed of it.

Part II
THE REDUCTIVE
ENTERPRISE

3

GUIDING VISIONS

THE HOPE OF ULTIMATE SIMPLICITY

GOING BACK to the distorting choice of stereotypes which makes it so hard for us to look at this whole problem, we probably need to start by examining further the very popular reductive option. Twentieth-century enthusiasm for reduction has been dictated, not just by a general policy of explanation through simplification, but by a faith that physical science can finally produce it. This faith has, of course, been hugely useful during the development of the modern sciences. But it has played a large part in raising the difficulties which are our subject. Indeed, throughout the twentieth century this approach has run into increasing troubles even in physics itself, and still graver ones elsewhere.

At the physical end, at the heart of science, the expected single explanatory structure did not emerge and it is not now expected to do so. It now seems that reality is not actually composed of a single set of ultimate 'building-blocks' at all. As Ilya Prigogine and Isabelle Stengers put it, the assumptions of classical science

> centered around the basic conviction that at some level *the world is simple* and is governed by time-reversible fundamental laws. Today this appears as an excessive simplification. We may compare it to reducing buildings to piles of bricks ... [The recent shift of scientists away from this perspective] is not the result of some arbitrary decision. In physics, it was forced upon us by new discoveries that no one could have foreseen. Who would have expected that most (and perhaps all) elementary particles would prove to be unstable? ... Quantum mechanics has given us the theoretical frame to describe the incessant transformations of particles into each other ... The models considered in classical physics seem to us to occur only in

limiting situations such as we can create by putting matter in a box and then waiting till it reaches equilibrium.[1]

Physics itself, in fact, no longer offers ultimate simplicity. This change naturally raises the question whether there is any longer a reason to consider physics as the sole route by which all other studies can achieve explanation. Accordingly, physicists themselves tend today not to be much interested in providing a reductive pattern that will clarify all the rest of thought. Though they do take their science to be in some sense fundamental, because it stands at the end of physical analysis, they no longer offer the simple, all-purpose, terminus of explanation which used to be hoped for. They tend rather to emphasize complexity.

As Prigogine and Stengers point out, this change makes possible a much less depressed view on the relation between the sciences and the humanities:

> Although Western science has stimulated an extremely fruitful dialogue between man and nature, some of its cultural consequences have been disastrous. The dichotomy between the 'two cultures' is to a large extent due to the conflict between the atemporal view of classical science and the time-oriented view that prevails in a large part of the social sciences and humanities. But in the last few decades, something very dramatic has been happening in science, something as unexpected as the birth of geometry or the grand vision of the cosmos as expressed in Newton's work . . . Science is rediscovering time . . .

> Traditionally, science has dealt with universals, humanities with particulars . . . For too long there appeared to be a conflict between what seemed to be eternal, to be out of time, and what was in time. We see now that there is a more subtle form of reality involving both time and eternity.[2]

THE PRICE OF PARSIMONY IN THE SOCIAL SCIENCES

When a physicist of Prigogine's distinction talks like this, the idea that all explanation can be safely directed towards the ultimate haven of physics starts to look less convincing. That offer, which was originally made by Descartes, has in fact been losing force throughout the twentieth century. But during that very time theorists more remote from physics have increasingly relied on it. The news that the offer no longer holds is very slow in filtering through to biologists and social scientists. The idea that rationality calls on us to reduce all phenomena to ultimate, unchangeable atoms – bricks or billiard balls ruled by a few simple forces – is still very strong in our culture. Thus, the embryologist Lewis Wolpert urges social scientists to increase their efforts to become 'scientific' in this sense, though he notes disapprovingly that they have not yet got far towards doing so:

> It is very hard to be reductionist in the social sciences. The ability to account for much of physiology and anatomy in terms of cellular behaviour, and then in turn to be able to explain cellular behaviour in molecular terms, as yet, has no effective equivalent in the social and psychological sciences . . . *It is not known what equivalent to the cell is required for understanding human behaviour, or even whether such an equivalent exists.* Psychoanalysis is much worse off than eighteenth-century embryology. (Emphasis mine)

But if there is no such equivalent, must not other kinds of explanation be looked for? No. Wolpert lays it down as a principle that no other approaches will do. Phenomena must not be referred to the larger wholes that are their contexts:

> Any philosophy that is at its core holistic must tend to be anti-science, because it precludes studying parts of a system separately – of [sic] isolating some parts and examining their behaviour without reference to everything else. If every process were dependent on its part in the whole then science could never have succeeded.[3]

This seems rather a wild claim, because the physical sciences constantly do take into account the relation of parts to wholes. Their explanations could not have got started had they not done so. For instance, in order to explain a leaf or a pine-needle, botanists naturally ask about the tree and the conditions under which trees live, as well as about its microscopic structure. The only reason why this area gets neglected is because so much is already known about it independently of science.

THE MEANING OF ATOMISM

However, many scientists, especially biologists, still agree with Wolpert in seeing the atomising approach as the only scientific form of explanation, and the word 'reductive' is sometimes used simply in this sense. Scientists often like this approach, partly because they are used to it and partly because they can use it to eliminate God by establishing determinism. The Greek Atomists and Lucretius gave atomism tremendous status by using it in this way to displace divine providence, and that function has continued to be central to its meaning. This symbolism is still very strong, so that talk of finding the secret of the universe by locating once and for all 'the basic building-blocks of matter' still provides a strong sense of metaphysical reassurance and a stick to beat the churches with, as well as the delusive hope of a solid terminus to theory.

This shift from moral to scientific status is something of the utmost importance. As Prigogine and Stengers point out:

> The urge to reduce the diversity of nature to a web of illusions has been present in Western thought since the time of the Greek atomists. Lucretius, following his masters Democritus and Epicurus, writes that the world is 'just' atoms and void and urges us to look for the hidden behind the obvious . . . Yet it is well known that the driving force behind the work of the Greek atomists was not to debase nature but to free men from fear, the fear of any supernatural being, of any order that would transcend that of men and nature . . . *Modern science transmuted this fundamentally ethical stance into what seemed to be an established truth*; and this truth, the reduction of nature to atoms and void, in turn gave rise to what Lenoble has called the 'anxiety of modern men'.[4] (Emphasis mine)

Atomism is one of many cases where a metaphysic that was originally invented to express a moral vision has ended by being seen as a simple scientific fact. Atomism did not recommend itself in the first place by any factual plausibility – indeed in its early forms it scarcely had any – but on moral grounds, because it struck people as a more honourable and decent world-view than a grovelling and superstitious terror of the gods. Since grovelling superstition is a persistent evil, this moral consideration has continued to recommend atomism, and has always given it, along with the deterministic and materialist structures that have grown up round it, notable moral prestige.

When, however, this whole structure became exalted to the position of sole arbiter of thought, far too much was expected of it and it then naturally revealed gaps which made it clearly inadequate for its role. In its turn, it too began to show seriously objectionable moral features, aspects which are no less obviously immoral than those of earlier world-views. For instance, taken literally, this structure requires fatalism. By evicting, not just the gods, but human and animal consciousness from the real world, it abandons that profound respect for life and for the experience of others which has always lain at the root of morality. By claiming to exclude from serious thought the immense structure of evaluative concepts by which we guide our actions, it positively discourages thought on practical questions. And by treating us as aliens in the universe that bore us, it has indeed generated despair and 'the anxiety of modern man'.[5]

It is quite normal that imaginative visions of the world should begin to display faults on this scale if they have become over-inflated and are allowed too much authority. But earlier world-views, when once they were seen to be immoral, have been suitably criticized and have often been amended. If they could not be sufficiently amended, they have at least been morally discredited. The peculiar and unlucky thing about the present situation is that people do not see the current world-view as open to this kind of challenge at all because they do not see it as one optional moral position among others. They perceive it as a scientific fact that must simply be accepted. It has acquired a quite unsuitable kind of authority.

Neither supporters nor opponents of this supposedly scientific world-view easily notice how far its metaphysical claims stretch beyond anything that can reasonably be called science, nor how much of science itself necessarily consists, not of facts, but of convenient assumptions which change from time to time. In this case, it is remarkable how the lapsing of these wider assumptions within science itself has failed to shift confidence in their scientific status.

Physicists, though they still use the word 'atom', have now quite dropped actual atomism. They no longer deal at all in hard, indivisible ultimate particles, but in forms of energy. Determinism, which has long been an extremely obscure concept, has now been dropped in some central areas of physics and is under question in others, such as chaos theory. And materialism, which may seem the most central notion of these three, loses its clear sense as soon as the idea of solid, inert matter is removed. No doubt current champions of materialism chiefly mean by it something negative – that there are *not* spiritual beings. This went well with a dogmatic behaviourist rejection of consciousness. But once the need to take consciousness seriously has been recognized, its meaning becomes very obscure.

The fact that those changes have hardly discredited these ways of thinking surely makes it clear that their status was not a scientific one in the first place. Essentially, they represented a moral stand. What was revered was an attitude that seemed courageous, indeed heroic, an attitude which, by striking soul out of Descartes' dualistic picture, opted for self-reliance rather than dependence on God. To accept this God-free metaphysic as being in some sense scientific was seen as itself an honourable and indeed a necessary act – a categorical imperative, a duty so strong that any confusions resulting from it must simply be accepted, and any considerations that conflicted with it must be overridden. This way of thinking is, I believe, still so strong at present that it makes any direct attempt to deal with the free-will problem impossible. Accordingly, though I have needed to say a little about predictability in this book, I have no confidence that the matter can be properly understood at all until the background presuppositions become clearer.

FOLK-PSYCHOLOGY?

In biology itself, this persistent faith in an outdated model does not necessarily do much harm, though it does bias enquiry away from many topics that need intelligent large-scale observation towards the more mathematical havens of molecular biology. But in the social sciences it is much more damaging. There, it has often grossly distorted complex social and moral patterns in order to support supposedly scientific reductive explanations. And as these social studies advance, the patterns they have to deal with grow constantly still more complex. Thus the two ends of the hoped-for explanatory process are steadily pulling apart to the point where the centre cannot hold.

What does it mean to be 'scientific'? By origin, the word is just a general one meaning 'methodical'. In recent times, however, it has been narrowed by rather unrealistic demands for a very specialized ideal of scientific procedure, an ideal which centres on imitating certain familiar models in physics. In discussing human conduct the ordinary, non-physical kinds of explanation which we use in common life tend to be dismissed as 'folk-psychology'.

This is a seriously misleading name. The word 'folk' cannot properly be used to refer to ideas that one uses oneself. It points to forms of thought belonging to other people, and rather quaint, unsophisticated people at that. But psychologists now use it for vital categories which all of us, including the smartest and most up-to-date of academics, must use every moment of the day and without which we could not carry on social life at all. These categories centre, as we shall shortly see, on the concept of agency, of people's personally doing things, rather than merely being part of an impersonal flow of events like clods in a landslide or pints of water in a river. Every professor of psychology who wonders casually what to do and how to do it, or asks what a colleague has been doing, or reads a whodunnit, or decides to go home, or holds anybody responsible for anything, is necessarily using these categories.

These ways of talking and thinking are not just quirks of language that science can correct. They are not like the idea that the sun actually goes up or down on the horizon. About the sun, we can indeed, with Copernicus's help, transcend that everyday language. We can, as Berkeley suggested, think about the sun in two ways, as rising and not rising. We can 'think with the learned and speak with the vulgar'.[6] But our respect for learning must not mislead us into thinking that the learned have got the whole truth. Vulgar speech has a point here. It rightly describes the subjective viewpoint, which is an essential aspect of the truth. We do indeed see something go up and down.

About the sun, that subjective viewpoint may be relatively unimportant. But about action, where what is really happening is so enormously closer to us than the sun, where we ourselves are actually the scene of it, that internal point of view is centrally important. The concept of agency which expresses it is something quite indispensable to us. Certainly that concept is obscure. We shall be grappling with its difficulties throughout this book. But the idea of getting rid of it makes no sense at all.

As Raymond Tallis puts it:

> The implication of the pejorative modifier 'folk' is that this framework is a primitive, unreflective one which will be superseded by a mature, scientific psychology rooted in neuroscience. Such a psychology will do without most, or indeed all, of those things – such as beliefs, thoughts and desires, even consciousness and awareness – that we normally attribute to people and which cause most of the paradoxes and problems that beset the philosophy and science of the mind. These bothersome entities will simply be eliminated.[7]

Everyday thought is seen as mere amateurism, a false start, casual guesses exploded by modern discoveries and useless for mapping social life scientifically. B.F. Skinner insisted that it is urgent to ignore this non-expert thinking, because even for practical needs we must have something more professional. 'We need to make vast changes in human behaviour.' For this purpose, he said, we cannot afford to

> carry on, as we have in the past, with what we have learned from personal experience, or from those collections of personal experience labelled history, or with the distillations of experience to be found in folk-wisdom and practical rules of thumb. These have been available for centuries, and all we have to show for them is the state of the world today.[8]

Skinner supposed that he was being scientific. But it is interesting to contrast Heisenberg's view about this:

> One of the most important features of the development and analysis of modern physics is the experience that the concepts of natural language, vaguely defined as they are, seem to be more stable in the expansion of knowledge than the precise terms of scientific language, derived as an idealization from only limited groups of phenomena ... One sees that – after the experience of modern physics – our attitude towards concepts like mind or the human soul or life or God will be different from that of the nineteenth century, because these concepts belong to the natural language and have therefore immediate connection with reality. It is true that we will also realise that these concepts are not well defined in the scientific sense and that their application may lead to various contradictions ... but still we know that they touch reality ... Even in the most precise part of science, in mathematics, we cannot avoid using concepts that involve contradictions.[9]

Skinner, however, implied that psychologists have ready a new and incomparably more effective scientific language, a fresh conceptual scheme that could do for human affairs what Galileo and Newton did for physics. By using it, they could thus also reform the world in practice.

Some – but not enough – later psychologists have stood back and put this over-confident but hugely seductive offer in a more realistic perspective. Thus, Rom Harré and his colleagues comment:

If we set aside commonsense psychology in the hope of developing a new and better theory of human thought, action and feeling from scratch, as it were, we must pay a heavy price. The cost is the danger of irrelevance . . . For most people, and for most purposes, commonsense psychology and its extensions, such as the psychology used in the law, is all that there is. It provides a complete working account of thought and action and defines the repertoire of our emotions. It represents the 'state of the art' to which scientific psychology may or may not succeed in adding.[10]

What common sense provides is then not amateur floundering, but an extremely sophisticated and highly developed repertoire, a huge indispensable range of relevant resources, though of course one that must still be attended to critically. To use an admirable image of Wittgenstein's, natural language and its concepts constitute the old city (Prague, say, or York or Edinburgh) to which certain neat, geometrically designed suburbs have lately been added. The proposal that those suburbs – the sciences – should be substituted for it, or should pull it down and redesign it in their image, is not very sensible.[11] As Harré and co. point out, the familiar relation between pure and applied uses of the sciences is not a proper parallel here:

The relation of commonsense psychology to scientific psychology is not the same as that between biology and farming, physics and navigation . . . or chemistry and cooking or dyeing. The commonsense component of each of these pairs was an atheoretical *practice*, a set of rules of thumb. The first step to science was the development of comprehensive *theories*. But commonsense psychology is itself a theory, or perhaps a cluster of related theories. It recognises a distinction between obsessional and voluntary action, and builds its explanations of what people say and do around the basic idea of action as the product of agents following rules and conventions to realise their intentions, plans and projects.[12]

Questions about individual agency and responsibility have always been central to human life. The concepts that have been evolved for discussing these things could not, therefore, possibly ever have been casual or sketchy. Nor could they have been remote from practice, like pre-Renaissance theories of physics. It is wild to compare the current state of thinking about motives with that uniquely dramatic moment in the history of science. But it is not surprising that the drama of that moment has gripped the public imagination so thoroughly that people often think of it as an essential part of the idea of science itself. 'Being scientific' can then seem necessarily to mean putting all existing thought on the bonfire in order to start afresh.

Of course the fact that we must still use common sense and common language does not mean that it is satisfactory. Of course the concepts we use in understanding motivation are always imperfect and always need constant further development. Of course it is true that, as Skinner said, they have not saved our civilization from reaching a grave crisis. But then neither has our enormous recent advance in all branches of knowledge done that.

It is indeed important to stress how alarming our present crisis is, and how much of it does seem to be due to faults and obscurities in our motivation. It is quite true that we do not sufficiently understand human aggression, nor, more generally, the extent to which human narrowness, meanness and short-sightedness manage to persist, leading to enormous waste of effort, undisturbed by high intelligence and copious information. And this lack of understanding is indeed a central element in our dangers today. But the hope of gaining that understanding through a clean slate and a sudden Newtonian revolution in psychology is no more plausible than any other nostrum that somebody might seek to recommend, solely on the grounds that it has not been tried yet. We have to start from what we have gained so far.

THE SCIENTIFIC IMPERATIVE

Since Skinner's day, psychologists have in some degree begun to recognize that all psychological discussion must make use of vernacular terms which start from ordinary forms of thought such as human agency. They have seen that even researchers carrying on the most rigorous experiments have to discuss them in these terms. Yet their attempts to move in this direction are still blocked by a terrible sense of guilt at being unscientific. Scientistic critics, who dismiss everyday psychological language on the ground that it would find no place in work done on a particle-accelerator, still cause alarm.

This 'scientific' imperative is not actually felt as an intellectual need based on a record of unfailing explanatory success. There is no such success. Instead, the need to be in some sense scientific presents itself more as a general moral principle, demanding a certain cold, incurious style and attitude. It calls for a kind of detachment which is not just a general commitment to truth, but a selective refusal to attend to truths about many striking aspects of the subject-matter. It is not empirical. It demands a willingness to ignore everyday facts which are not yet paraphrased and represented in theory.

It does not aim primarily at providing a useful instrument for understanding just those problems that are, at the time, pressing on us and needing to be dealt with, but at developing that theory itself towards formal completeness – towards integration into an ultimate Theory of Everything. The fact that physicists now use this name – Theory of Everything – to describe their current project of bridging gaps within physics itself is highly significant. Such a name implies that a satisfactory ordering of physics cannot fail to trickle down and produce order in everything else.

IS REDUCTIVISM NEUTRAL?

Am I exaggerating here? Is there really this peculiar style and temper? Are such questions value-loaded? Scientists and philosophers who think of reduction as a purely objective, formal aspect of method are often surprised to be told that they are taking part in a drama. They point to cases such as the reduction of chemistry to physics. Surely, they say, this is just an innocuous piece of translation, uncontroversial and without moral consequences?

Reductions like this one might indeed not have caused alarm if they had stood alone. 'Methodological reduction', the mere translation of one set of terms into another and the use of the second to clarify the first, can be quite innocuous.[13] There is, however, always something significant about the fact that it does not work both ways. Reduction is not supposed to be a symmetrical relation, and when the model is exported to widely varying studies, this lopsidedness can become very important.

In any case, this kind of harmless methodological reduction requires a special, quite rare, kind of relation between the studies involved – a relation which perhaps is fully present only in the instance of chemistry and physics. These disciplines have grown up together with related aims. They are both highly abstract systems. Neither system has much direct, general implication for common life, so the relation between them is indeed mainly one between their forms.

Even biology, however, is by no means so closely linked to chemistry as chemistry is to physics. Important and fascinating though the relation between them is, attempts to press it into a similar subjugation have repeatedly led to serious distortion, as Arthur Peacocke points out.[14] Francis Crick, himself a vigorous habitual reducer, interestingly reinforces this point. He remarks that, because evolution has not been an elegant, planned process but an age-long, opportunistic source of clutter and complication, the biosphere often resists attempts to order it on neat and simple patterns:

> While Occam's razor is a useful tool in the physical sciences, it can be a very dangerous implement in biology. It is thus very rash to use simplicity and elegance as a guide in biological research ... All this may make it very difficult for physicists to adapt to most biological research. Physicists are all too apt to look for the wrong sort of generalizations, to concoct theoretical models that are too neat, too powerful and too clean.[15]

It is, of course, always more fun reducing than being reduced. Not surprisingly, Crick finds the over-simplicity of the physicists' view of his own subject much more obvious than his own over-simplicity in approaching the social sciences and humanities. Theoretically, however, he sees both processes equally as merely ways of translating between various studies, and therefore as being value-free.

WHY PROPAGANDIST REDUCTION WORKS

Notoriously, however, there exists also another, much more familiar kind of reductive talk that certainly is value-loaded. It is the kind that brings together two aspects of life which are both of great practical consequence and tells us that one of them does not matter because it is really only a minor form of the other. That is what Jeremy Bentham was doing when he proclaimed that justice and obligation could be fully explained by the principle of utility, which reduced them to their consequences for pleasure and pain:

> Systems which attempt to question [the principle of utility] deal in sounds instead of senses, in opinion instead of reason, in darkness instead of light ... When thus interpreted [as expressions of that principle], the words *ought*, and *right* and

wrong, and others of that stamp, have a meaning; when otherwise, they have none.[16]

Bentham thus made his one-sided moral position look plausible in
a way which Utilitarians have often continued to use – namely, by
making it look like a combination of a factual statement and a claim
to logical necessity. He implied that the facts are so simple that there
is only one intelligible way of describing them. Anyone who fails to
see this must therefore be simply stupid. The logical mistakes of
Bentham's tactic have been noted often enough, but the reason why
it remains so persuasive has not, I think, been fully understood.

Bentham's view is really a value-judgement, an assessment of the
importance of pain and pleasure among other values. But it does not
get judged and defended in the appropriate way by relevant
arguments about those other values. Instead, it simply knocks down
its opponents by its supposed authority as a piece of factual
knowledge. In an age when the prestige of factual knowledge –
especially scientific knowledge – has steadily increased, this tactic has
often been a winner. Psychological hedonism and egoism are still very
powerful. The trouble with such views is not just that they commit
the 'naturalistic fallacy' by deriving value-judgements from facts. It is
that they make it look as if the value-judgements have vanished, being
simply absorbed into the factual ones. We have left the realm of
awkward debate (morals) and arrived in that of certainty (science).

In his very different style, Nietzsche, too, often used this sort of
dramatic, ideological reduction:

> This world is the will to power and nothing beside . . . Life
> itself is *essentially* appropriation, injury, subjugation of the
> strange and the weaker, suppression, severity, imposition of its
> own forms, incorporation and, at the least and mildest,
> exploitation.[17]

This is certainly not a system of translation between two studies, nor
indeed a reduction of mind to body. It is primarily a psychological
reduction, a proposal for a new view of human motives, though it
is somewhat casually extended to cover non-human life as well.
Among these human motives, it 'reduces' all other apparent varia-
tions to a single chosen model, encouraging us not to bother about
them by flattening out their distinctive properties because they are
not important. It generalizes a chosen insight sweepingly, to make
it cover the whole psychological scene.

Is it perhaps a mistake to call these two diverse-looking approaches both 'reductive'? Are they really connected at all? Perhaps in principle they might not have been, but our history has linked them at a deep level. Both Nietzsche's and Bentham's claims do indeed differ in striking ways from the reduction of chemistry to physics. They differ in subject-matter and also in form. Both of them – for both are essentially simplifications of motive – gather many motives together under a single heading rather than analysing each of them into constituent particles. But the form of this approach still has a crucial common feature with scientific reduction. Each of them puts forward a single interpretative scheme as having the sole authority to explain all that needs explaining in the area that is 'reduced'. Each of them claims an exclusive direct line to a single fundamental level of reality underlying currently observed appearances. Each of them does, therefore, flatten out the differences between other possible approaches by dismissing them all in favour of its chosen method.

This monopolistic claim is certainly a strange one, and it has huge consequences. In any field of thought that has not already been processed to abstraction, all sorts of questions normally arise that need answers from outside that field. They are answered by outside consultation – by bringing in whatever conceptual resources suit their form and subject-matter. Thus for instance historians, in studying history, must constantly raise questions involving law, geography, linguistics, agriculture, engineering, economics, medicine and a hundred other disciplines. To answer these questions, they know they must seek out the appropriate conceptual scheme and often consult the appropriate outside specialist. *They do not assume that the sole reality underlying history is a single, vast, hidden process, a process formally simple and accessible only by a single privileged thought-pattern*. There have indeed been theories of history which have assumed this, such as the Marxist one. But they have turned out to be very bad ones.

THE USES OF WONDER

This claim by one discipline to exclusive mining rights for a whole area seems to be the source of that devaluing that we usually associate with the word 'reductive'. When we say that any actual thing in the world (as opposed to a concept that is already abstracted) is quite simple and needs only one sort of explanation, we are, almost unavoidably, saying that it is something fairly trivial. Spoons are a great deal easier to explain than laws or trees or earthquakes or passions or symphonies, and even spoons have several aspects – culinary, metallurgical, aesthetic and what not. Anything more important than spoons is bound to have many more.

Important things are, by definition, ones that have many connections and many aspects. Certainly we can sometimes praise people or their actions for being simple. But that seems to be because, in their particular situations, a certain particular kind of simplicity is called for. Explaining why it is called for can be very complicated indeed.

In general, important and valuable things seem to be complex ones which provoke wonder. These things impress us, producing a sense that there is a great deal about them that we do not know and perhaps do not even know how to ask, and that we are not likely ever to get to the end of pursuing it. As Kant put it, discussing 'the animating principle' in art, they are things that 'induce much thought, yet without the possibility of any definite thought whatever, i.e. *concept*, being adequate to them'.[18]

Champions of the intellect (such as the Stoics) have worried about wonder because it seems to involve welcoming a kind of indefiniteness, a sense of the slightness of our knowledge. Surely, they say, we ought not positively to revel in our ignorance? Ought we not to aim steadily at knowing everything, at penetrating all mysteries and so at being no longer surprised at anything? Ought not wonder – the discovery of our ignorance – to cause us merely irritation rather than this mindless delight that we call admiration?

There is surely a confusion here about the kind of beings that we are. We do not seem to be creatures for whom the idea of knowing everything, or even of fully understanding anything, makes a lot of sense. For one thing, this idea would mean that there was only a certain, finite set of questions that could be asked. But we know that it is in the nature of life that, for every question we succeed in answering, ten more start up, many of them of quite new kinds.

Normally, wonder does set us to work on answering some of these new questions. But before that can happen, wonder has a more static, contemplative phase in which we simply sit back and pant, taking in what is being revealed. Without that phase, our thought cannot go on properly to generate its further questions. And in any case, must not the aim of our thinking be this positive enjoyment of the knowledge we have got rather than the perpetual scurrying on to dig up something further?[19]

Reductive explanation, however, inhibits both of these phases. It tells us that the answers we already have on this matter are quite adequate. The subject has no further complications and does not merit further attention. Now there are obviously cases where this is true – namely, where we are thinking about something that actually is simple or trivial or misleading. Reduction can then usefully redirect our attention away from slight or unprofitable topics to important ones, from spoons to laws, and also from merely imaginary problems to real simplicities. But if it is to do this, it has to be carefully directed at a well-chosen target. Its destructive effect has to be fully understood and used on what needs destroying. Its negative aspect – the exclusion of other forms of thought – has to be justified. And the justification of this will be different in each case. What really cannot work is a single, all-purpose reduction of all subjects to a single substrate.

4

HOPES OF SIMPLICITY
———— •◆• ————

THE ILLUSION OF
THE SINGLE VIEWPOINT

THE CORE reductive mistake is, then, the idea of a single funda-mental explanation. This idea of the fundamental is a very powerful metaphor. Its working image is that of a pile of semi-transparent films, a pile whose top members are relatively flimsy and the lower ones stronger, over a solid base which is the real object of study. All the layers display patterns which represents the basic reality to some extent, and it is admitted that sometimes the upper ones may often do so well enough for everyday purposes. But these are mere convenient appearances. Enquirers can only hope to move towards real knowledge by disregarding the upper layers and piercing steadily down towards the solid truth at the bottom of the pile. To remain at the upper levels is frivolous.

This is a most misleading pattern for understanding the relation between different studies. The image that is actually needed here is something much more like the familiar one of viewpoints. The object that we are studying – for instance human behaviour – needs to be conceived as something vast, complex and relatively distant, perhaps something like a mountain. Different approaches show this mountain from different sides. The various views of it can differ dramatically, so much so that sometimes the enquirer wonders 'Can this be the same mountain? Can these two sets of people be inves-tigating the same phenomenon?' But patient travelling between the observation stations will show why these differences arise.

In particular: questions about how people act and how they are moved to act are far more complex than we usually notice. We have two main approaches to these questions, approaches which we constantly need to compare. One approach uses the pattern of regular causal sequences. This approach is largely formulated in the physical and social sciences. The other works in terms of the kind of concepts that the people who are acting themselves use, concepts that are helpful in deciding how to act. This second approach is formulated in terms of reasons for acting. History makes use of both approaches.

On neither side can the formulation be anything like complete, nor does it seem likely that the relation between the two sides can ever be made fully clear. Yet in any given case, data supplied from both sides are so continually used together that a highly usable working relation evolves. We can say a great deal about how the two contexts are connected. It is essential, however, that we should never become so obsessed with one kind of approach as to forget about the other. We must always keep in touch with what is going on on the other sides of the mountain.

OUGHT WE TO EXPECT OMNISCIENCE?

That image concedes that we shall never know everything about the matters we are enquiring into, any more than we could penetrate every detail of a mountain. Human affairs are in fact rather complicated and our powers of understanding are limited. This last point is important. Whatever else evolution may have accomplished, it plainly has never put any pressures on our species that would be likely to give us faculties capable of grasping this or any other subject-matter *completely*. Indeed, it is hard to imagine what kind of evolutionary pressures could possibly do this, whether on our own species or any other.

This is a point that ought to be particularly clearly seen by those dogmatic neo-Darwinists who are today most sure that every development in evolution has been a response to a particular form of selection and must therefore serve a definite function. Evolutionarily speaking, it would be quite extraordinary if we had faculties capable of finding a universal explanation. Subject-matters are solid; there are narrow limits to what we can know about them. So, though (as I have said)[1] reason may in some sense demand that we assume that everything does make sense and have an explanation somewhere, reason certainly does not call on us to assume we can always find it.

This admission that our powers are incomplete seems to be a concession that should be made in any case, perhaps even one necessary for sanity. It surely blocks the ambitious assumption of a single, discoverable, universal form of explanation that has been central to rationalist philosophers and now is so to modern reducers. But it does not condemn us to mere despairing, relativistic confusion. We can walk round and see how the viewpoints are connected – that is, we can investigate the various enquiries themselves and try to relate them. Sometimes, indeed, there is special reason to give priority to one particular method because its view is the one most needed for the question we happen to be asking. But in general, the job of enquiry is not to set up a competition and to choose one view as the true one. Instead, it is to build up a composite picture from them all. (We will ask more fully what this means later in considering the case of Evan Jones in Chapter 6 of this book.)

In the case of physics and chemistry a monopolistic, one-way explanatory relation is quite reasonable. Because chemistry is already such an abstract study, the kinds of explanation that can be needed in it are sharply limited. Economic, legal or linguistic questions simply do not arise there. Modern chemistry is – for good reasons – conceived in a way that already concentrates attention on physical explanations. This pattern has made it extremely successful, without having any obvious moral consequences.

Even the case of biology is more complex however, because biology raises far more kinds of question than chemistry. Francis Crick's claim that 'the ultimate aim of the modern movement in biology is in fact to explain all biology in terms of physics and chemistry' is a controversial one because it ignores this complexity.[2] Though it is not directly political, this claim is certainly a piece of academic imperialism, a claim about which questions biology should attend to and therefore about how it should develop. It is not, then, neutral or value-free. And when we turn to a topic like motivation, the notion of neutrality ceases to look plausible at all.

THE IMPORTANCE OF SELECTION

Psychological reductions such as Nietzsche's are clearly not designed as neutral pieces of explanation. They are not ways of resolving a formal problem but explicit pieces of propaganda. They are calls to look at life in a different way and to live it differently. Between them and the unexciting case of chemistry lies a whole spectrum of reductions varying widely in spirit and intention, but always licensing one interpretative scheme to dominate all the rest. When the subject-matter is already formally limited, this can be in order. But when it is something large, sprawling, indeterminate and of general human concern, such moves are wildly inappropriate.

It is not easy to see how this habit of simplifying complex subject-matters wholesale, by an act of faith, could ever be justified. What is clear, however, is that it cannot be justified on the ground that it is morally neutral. A choice of interpretative scheme is not just a choice of convenient laboratory equipment. Different interpretations express different emphases. They endorse different principles of selection. They determine what will be attended to.

Claims to a monopoly of explanation therefore unavoidably deal in values and often say something very drastic about them. They involve supporting one general attitude to life against others. Yet increasingly, in the last century, they have come to be seen as somehow objective and 'scientific', solid products of science, impartial theories having the authority of proven fact. Nietzsche himself did not often press this claim to scientific status. But both Marx and Freud did, and it has become increasingly important to their successors.

IDEOLOGY MATTERS

I have suggested that, in discussing reductivism, we need to attend seriously to popular, openly ideological writings such as those of B.F. Skinner and the campaigning sociobiologists, and that many philosophers today refuse to do this. Much philosophy has indeed become so specialized during this century that people working on big metaphysical problems are quite liable to think that ideologies exist only in order to provide examples of formal confusion for the logic class, being otherwise left to the politics department, or at worst allowed to intrude occasionally in moral philosophy, which has itself been ghettoized in an enclosure down the corridor.

This shift away from the tradition of the great European philosophers is not just a change of fashion or of teaching practices. It expresses a solid philosophical mistake, a misguided attempt to divide the forms of thought from their function. Conceptual schemes that are used to discuss large issues like free will or the mind–body problem can never be tidied up on purely formal principles. The question about them is never just whether they are consistent, but what they are consistent *with* – what background presuppositions, what wider imaginative vision has gone to shape them.

These concepts have always grown out of practical disputes. They are always to some extent adapted to settle those disputes one way or the other. Certainly it is sometimes possible to detach such concepts from their original use deliberately by argument. Alternatively, it is also possible to declare that one actually approves the underlying attitude, and to give one's moral reasons for approving of it. But neither of these things can be done without stating these background issues openly.

It is often, therefore, necessary to examine vulgar but powerful ideologies, making explicit the unexpressed background that gives them their power. This kind of examination has, in fact, always been one of the great uses of philosophy. The recent revival of 'applied philosophy' seems to signal some welcome return towards recognizing this function. But the isolationist idea that metaphysics ought to be done as a quite neutral subject, in abstraction from its guiding world-pictures, still prospers. To add one more image to those previously suggested, it works much like an attempt to deal with a pile of mixed jigsaw-pieces, coming from different puzzles, by putting together a set that makes a tidy sky, and supposing that this will still leave us impartial, not committing us to one eventual picture rather than another.

MIND–BODY PROBLEMS

This kind of hopeful purism is all the more surprising when we consider how strong and obvious are the forces making for partiality here. Everybody with experience of academic life knows how easily the notion of a single dominant explanatory scheme can merge into ordinary academic imperialism. It is also very easy for people who have found an intellectual scheme which fits their thinking to feel sure that it must be the only right one. Important, influential reductions share both these attractions. Among these successful imperialistic conceptual schemes, the systematic, general reduction of mind to body has held a central place.

This has not just involved the purely formal, methodological suggestion that mental phenomena might be causally explained by reference to physical ones. That path of enquiry has of course been pursued, often usefully. But the kind of 'materialist' reduction that seemed to have obvious implications for moral and political causes has naturally been noisier and more influential. Here reductivism became cruder and more dogmatic. Tentative, useful ideas about how things might best be explained harden into aggressive cosmic doctrines about how things fundamentally *are*, how, for instance, minds are less real than bodies, or indeed are not real at all. Epistemology gave way to ontology.

The campaigns for which the resultant doctrines have been used have varied, and their direction is now changing significantly. During the Enlightenment they were straightforwardly political. Hobbes and other Enlightenment materialists forged their metaphysic essentially as a weapon against the political power of the churches, which was commonly linked with that of the State. Anti-clericalism determined the kind of materialism that was favoured. But today the churches have little political power and are not at all a central issue. Although mutual misunderstandings about 'God and science' continue to be aired, they are actually a distraction from the real trouble.

EXPERTS VERSUS FOLK

Modern materialism has always had another aspect which seems now to be becoming central to it. It attacks not God but subjective experience. It now tends chiefly to target 'folk-psychology' – meaning the everyday, vernacular notions of the human self by which we normally live, including, of course, our notions of freedom. As we have seen, reducers offer to substitute for these ideas more scientific conceptions provided by the appropriate experts, who are, of course, physical and social scientists.

In the behaviourist epoch, it was thought that this could best be done by simply ignoring consciousness, which was held either to be – as J.B. Watson suggested – actually unreal, or, as Skinner preferred, an inert extra, a mere froth that did not affect the real action. Since this proposal proved hard to explain, consciousness has now been promoted instead to the status of a Problem, a muddled topic which science is willing to take over once it gets the appropriate research money. It is noticed, however, that not everybody is hopeful about this project. As Lewis Wolpert remarks, some people argue

> that human behaviour and thought will never yield to the sort of explanations that are so successful in the physical and biological sciences. To try to reduce consciousness to physics or molecular biology is, it is claimed, simply impossible. This claim is without foundation, for we just do not know what we do not know and hence what the future will bring ... A characteristic feature of science is that one often cannot make progress in one field until there has been sufficient progress in a related area. The recent advances in understanding cancer were absolutely dependent on progress in molecular biology.[3]

Science, in fact, will get there in the end if it only goes on long enough. But suppose it is going in the wrong direction? No amount of expensive travel westwards will get you to the south, nor is a microscope much use when you are trying to see a mountain. The methods that an enquiry should use depend on the kind of question that it is trying to answer. Wolpert wants to put the burden of proof on those who advise using different methods for the distinctive kinds of question that arise about conscious subjects. But clearly that burden must rather fall on those who claim that narrowly conceived 'scientific' methods – which have been carefully devised to exclude all questions of this kind – can now be stretched to cover them.

Microscopes, those splendid tools of modern scientific method, are also its most significant symbol. Microscopes reveal new patterns, patterns which can sometimes be of the utmost importance. But they make the original macroscopic phenomena invisible. When we want the facts at that everyday level, we have to put away our microscopes. If (for instance) the problem is why certain people are anaemic, we must answer questions about their way of living as well as ones about the constitution of their blood. For many of the most relevant of those questions, neither the microscope nor the scientific method that it serves is any use at all. (For instance: are these people happy? Are they fairly treated? How are they trying to live?) Different patterns, different ways of thinking must be brought in. Similarly, sand is a wonderfully versatile building-material, but it can't be used as a substitute for coffee.

The supposed new understanding of the self which is supposed to emerge from this missionary project on consciousness has not yet got near enough to the drawing-board to be usefully discussed. Until the vast confusion on this front has been dealt with, it is not likely that the God-and-Science debates can make progress either. A reductive technique which cannot find a useful language for discussing even our most familiar, everyday experiences is not likely to be able to say much about the more puzzling ranges of human thought, such as those involved in religion.

PSYCHIATRIC TROUBLES

This kind of reductive materialism prevailed strongly in that most metaphysically dogmatic of states, Soviet Russia. On principle, throughout the Soviet empire, no conception of mental disease was admitted and no treatment of it practised that was not strictly physical. This same metaphysical view is, however, still no less influential in the West. It provides a striking example of how deeply these apparently theoretical questions can affect common life. In psychiatry, the reduction of mind to body is now seen as a major factor in determining diagnoses and methods of treatment. As two concerned practitioners in this field have put it,

> Despite the ambiguity and complexity of psychiatry, it is striking that many students begin its study with the appearance of having solved its greatest mysteries. *They declare themselves champions of the mind or defenders of the brain* . . . Psychiatrists, after all, are the only physicians regularly said to have this or that 'orientation', and the labels mark them as friend or foe . . . In order not to be wounded before they even know what the fight is about, beginning students may seek protection in one or other warring camp . . . The unfortunate result is that many of them become partisans – and needless casualties – in denominational conflicts that have gone on for generations and that they scarcely understand.[4] (Emphasis mine)

As the authors point out, this metaphysical issue cannot be ignored. It is not just a trivial divergence, a mere preference for different language or imagery. It is

> more than a question of taste whether we think about schizophrenia as a clinical syndrome . . . as a set of maladaptive behaviours, a cluster of bad habits that must be unlearned, or as an 'alternative life style', the understandable response of a sensitive person to an 'insane' family or culture.

> Each of these proposals makes different assumptions about the phenomenal world and its disorders, and each has different consequences for psychiatric practice and research . . . The result of ignoring the fundamental differences between perspectives is not to diminish sectarianism but, in the end, to encourage it.[5]

Their quite long list, only part of which is quoted here, of possible ways in which schizophrenia might be seen shows plainly what tends to be wrong with reductive, exclusive approaches to large-scale problems. All these suggestions seem clearly worth taking seriously. One might reasonably expect that even the wildest of them might play some part in a proper understanding of this very obscure complaint. It seems reasonable to suggest that they would best be seen as viewpoints belonging to sets of investigators encamped round the mountain of mental trouble. Yet the temptation to choose one and to take sides rigidly is extremely strong for a profession that feels the 'scientific' imperative compelling it to choose only one approach.

Modern reducers often do not seem to be aware of these practical consequences of dogmatizing. They tend to see themselves primarily as conceptual clarifiers, imposing the only possible consistent order on theories. The reductive temper is, I suppose, typically one which confidently hopes to short-cut these large, complex and painful questions by a single Damoclean stroke of the scientific sword. It sees strokes of this kind as purely formal, neutral moves, internal to a particular physical science or to a branch of philosophy.

Such strokes, however, quite evidently cut far beyond the boundaries of any science. And when they are used – as they are here – to justify radically different practices, it is natural to conclude that moral and ideological motives for using them are involved as well as the theoretical ones. Excluding these motives would be an extraordinarily difficult project. Even when people try to drive drama out of such discussions with a pitchfork, it constantly returns. At the level of the school debating society, one can constantly hear remarks like, 'When you get right down to it, a human body is just five pounds worth of chemicals.' Reducers of this kind may be simple-minded, but the tradition of European thought has given them far too many precedents for being so.

5

CRUSADES, LEGITIMATE
AND OTHERWISE

PSYCHOLOGICAL REDUCTIONS

As we have seen, propagandist reduction is quite old, and it has often been very influential. Thus, when Hobbes wrote that 'no man giveth but with intention of Good to himself, because Gift is voluntary, and of all voluntary Acts the Object is to every man his own good', he was not just clarifying a convenient system of definitions. He was recommending a new interpretative habit which would make people more self-regarding and stop them consenting to wars of religion. This particular definition was merely one more nail for the coffin of hierarchical, feudal thought-patterns.

The same is true of Freud's many loaded definitions, for instance 'Parental love, which is so moving and at bottom so childish, is nothing but the parents' narcissism born again.'[1] Freud wanted to alter people's whole attitude to their feelings in a way that would give them better control over their emotional lives. Both his aim and Hobbes's were reputable. Nor is there is anything wrong with using striking paradoxes to promote them. What makes such moves noxious is a determined claim to exclude all other ways of thinking. What makes them disreputable is concealment. If reducers hide their propagandist aim, if they pretend that these loaded paradoxes are mere impartial scientific conclusions, then reduction is being used illicitly. The atmosphere of science is then being imported to give unjustified authority to a piece of propaganda.

Writers like Hobbes and Nietzsche did not conceal their aims in this way. They clearly were not putting forward either formal, logical clarifications or factual statements. They were recommending general ways of interpreting facts. Such proposals for interpretation do have a connection with the facts, but it is an evaluative one, telling us that certain ranges of facts are more important than others. They can also alter our willingness to believe certain kinds of evidence. Their root is the moral judgement that certain facts *ought* to be attended to rather than others. In Nietzsche's case, the claim that there is 'nothing beside' does not mean that rival factual suggestions have been disproved, or disqualified because of some formal fault. It means that rival approaches to life as a whole have been discredited on moral grounds as cowardly and unvirile.

This tradition of psychological reduction, though still active, carries much less clout today than physicalist reduction does. It no longer seems so 'scientific'. The reduction of mind to body – the metaphysical insistence that description in terms of the physical sciences is the only really proper kind of description – is now seen as more important, indeed it now often seems to have the peculiar status of being seen as being itself an established fact of science. In the confusing welter of present-day life, this apparent certainty and finality has enormous appeal, an appeal which no doubt does much to explain the popularity of reduction. That is why the meaning of various reductive moves has to form a very important part of our subject-matter throughout this book.

THE PLACE OF FAITH IN RATIONALITY

The difficulty is that, since the seventeenth century, our tradition has insisted on a peculiarly high standard of certainty that can supposedly be found only in science. But such certainty as is available to us at all is in fact mainly found elsewhere. All human enterprises, including the sciences, constantly depend on ways of thinking which can in no way be reduced to scientific methods. For instance, we trust, and have no choice but to trust, most of the evidence of our own senses and memory, and of the reasoning-powers by which we assess them. We also trust most of the utterances of those around us. If we did not, we would not have that general knowledge of the world from which science starts, and we certainly could not use other people's testimony as scientific evidence. Certainly we can question particular suspect data, both from other people and from our own faculties. But we can only do this by using the mass of other unsuspected data as a standard.

In order to trust people in this way, we have also to credit them with an inner life comparable with our own. If we did not – if we thought that they were machines or dream-figures – we could not treat them as reliable witnesses. The reality of these 'other minds' is not just a hypothesis confirmed by experience. It is a vital, innate assumption without which speech would be impossible, no hypotheses could be formed, and distinctively human experience could not take shape at all. We assume, too, with astonishing confidence, that the future will continue to run on roughly the same lines as the past. And in general we also trust the value-judgements by which we guide our actions.

It is not possible to treat these forms of trust as irrational. They are necessary preconditions for reasoning itself. Faith in these things is as necessary for thought as it is for life. Somebody incapable of this faith would not rank as a specially perceptive, critical thinker, but simply as being autistic or insane. Yet these assumptions are much too large to be established by any science. Reductionist theorists often do not notice this. Thus Lewis Wolpert concedes that

> consistency and universality in the laws governing nature are basic, and usually unstated, assumptions that scientists make. *But such assumptions are testable*.[2] (Emphasis mine)

How, then, would we test them? How would we set about checking whether the whole vast mass of facts that we have not yet experienced – facts in the past, present and future – follow the same patterns as the tiny sample of which we have records? And again, how could we check whether those records themselves are reliable? Assumptions like these are not just provisional approximations which will be replaced one day by more refined scientific proofs. They are the ground from which our sciences, along with all other human activities, start. They form the central framework both of our natural equipment for thinking and of the social faculties by which we handle our own and each other's daily experience. They are not something that could possibly be bypassed.

FREEDOM IS NOT DETACHMENT

Something very important follows here for our notions about freedom. If this basic trust in the world and those about us does not make us unfree – if it is not a weakness, but rather a strength for us to acknowledge our continuity with our surroundings in this way – then the freedom that we need is not to be sought in detachment, in isolating ourselves from the rest of creation. It has to lie rather in taking our proper place within it, in rightly understanding our relation to it. We will come back to this point in Part IV.

Meanwhile, as far as the status of the sciences is concerned, it is important to see that this unavoidable reliance of all our thought on faith in extra-scientific elements is not a tragedy. It simply shows that the physical sciences are like microscopes. They are purpose-built tools, designed with the utmost skill for use on certain special kinds of subject-matter and shaped to handle a certain chosen level of abstraction. It is just this limitation, this deliberate narrowness, that has made them so successful. It was precisely in order to guard this narrowness that Galileo, Newton and the other architects of modern science carefully excluded from its range most topics of direct importance to human life, topics such as purpose, and also subjective pain and pleasure. That is why its methods cannot now handle these topics. Yet of course all of us, including the most fanatical reductionist, must continually attend to such topics in our thinking.

DESCRIPTIONS DO NOT COMPETE

This, too, is why there cannot be only one right and proper kind of description – namely the ultimate 'scientific' one. The debating-society example of describing a human body as five pounds worth of chemicals shows the difficulty. If the officials who are supposed to provide bodies for dissection at the medical school decide just to send in crates of chemicals, there will be trouble. However carefully they weigh these crates and however sure they may be that social and biological descriptions are only superficial and provisional approximations, they will not have sent what was ordered. The ordinary social description is the definitively right one here. It is not just a convenient device for use at a crude level. There is no sense at all in which it is provisional. There is no 'bottom line', no series of levels of reality towards whose end all descriptions aspire and by which they can be faulted.

The proposition that a human body is just five pounds' worth of chemicals only makes sense if we add the words 'chemically speaking'. But of course this makes it look much less exciting. The claim to exclusive status, which is the exciting part, is also the hollow one. Different kinds of description do different kinds of work. Even when they refer to the same item, they can say quite different things about it and can have totally different consequences. These consequences can indeed sometimes clash, and must then somehow be reconciled. But *there is no general reason to expect one description to be the only right one and to reduce all others to it.*

For instance: if a particular tree is described by a botanist, a forester, a carpenter, a landscape painter and a person who has always lived under it, these very different descriptions do not have to be placed in hierarchical order. They do different jobs and supplement each other. They are not in competition. None of the describers needs to be a physicist, nor need a physicist's description be added unless physical questions arise. Explanation is the answering of questions, and questions can be of many different kinds. They can arise out of different doubts, and the points most relevant to those doubts are the ones on which explanation ought to concentrate.

Though (then) rationality does sometimes call for simplification, it gives Procrustes no general licence to iron out all such complexities. Reason is not a simple tool for simplifying, nor is rational thought an intellectual monoculture. And this (again) is not a tragedy. The world does not relapse into helpless confusion just because things have more than one aspect and can be correctly described in more than one way. On the contrary, overlapping pictures taken from different angles provide the right way to get a reasonably unified notion of an object. Grasping this benign plurality is the first step in a rational approach to language.

THE PITFALLS OF
PSYCHOLOGICAL ATOMISM

The legitimate point of reduction is explanation. Its officially declared aims are clarity and parsimony. In the physical sciences, its most familiar form is the explanation of wholes by analysing them into parts, on the model of splitting living organisms into cells, cells into their chemical elements and chemical elements into their physical particles. As we have seen, these methods have had enormous success because they provided the right approach for a particular crucial problem – namely, finding what stuff things were made of and how that stuff worked. In these cases that problem was indeed the outstanding one, the obstacle which was holding up numberless other paths of enquiry. It was the right door to try, and the key provided did prove to open it. Proper, intelligible answers to these atomizing questions did exist, and were found invaluable for many other kinds of purpose.

It is no wonder, then, that this success produced the usual tendency to claim an exclusive status, making this the only kind of explanation that could be viewed as 'scientific' at all. In particular, it is not surprising that the modern age's search for certainty led to attempts to extend this atomizing method to psychology by finding the ultimate units of conduct. This approach has produced a series of different 'atomizing' accounts.

This attempt, as Lewis Wolpert rightly complained, has had no success yet and shows no signs that it ever will. The most recent of these projects has been the behaviourist analysis of human conduct into a set of unitary physical behaviour-patterns, each supposedly made up of a predictable series of reflex responses to particular stimuli – a project which has now been abandoned as useless. More persistently, there has also been 'social atomism', the principle that group behaviour can only be understood as the sum of particular acts by individuals; 'there is no such thing as society'; 'the state is only a logical construction out of its members'.

The misplaced ideal of science is one source of that decision, but, as Rom Harré and his colleagues remark, other, more ideological forces have also contributed to it:

> Much contemporary social psychology is strongly individualistic. This seems to be the result, not of the nature of the subject-matter, but of two main features of American social life which are reflected in the methodology of social psychology through the recent dominance of the United States in this field. The deep-seated individualism of American culture makes it very difficult indeed for American scientists to conceive of genuine *group* activity. Through the influence partly of individualism and partly of the prevailing 'technologism' of contemporary American culture, an approach called the 'experimental method' has been adopted. Each 'subject' is studied separately.[3]

As they point out, this approach distorts social psychology hopelessly, because in social action contexts are all-important. In any case, choices of method like these are not value-free. Where the subject-matter is so complicated, any choice of a particular unit of description is also a choice of what to concentrate on, and therefore of what to treat as important. There is no such thing as a single, neutral, 'scientific' way of conceiving and describing behaviour. The hope of dividing that behaviour into smaller particles does not furnish such a way. This attempt is bound to be tendentious, and is also often a mere recipe for obscurity. Unless we have a grasp of what a person or animal is trying to do, it is often not possible even to find suitable names for the movements of its limbs. And if the activity is a social one, we also need to grasp the larger social context which gives sense to it before we can intelligibly describe it.

This need is built into the nature of language. Speech has been evolved for describing the world that people normally experience. Its structure is not just a trivial, arbitrary addition to that experience, but an indicator of its basic shape. Attempts to invent technical terminology unrelated to that experience only produce unintelligible jargon. Though languages vary widely, they never work atomistically in the way required here. In seeing and describing an action, people constantly need to take account of the intention behind it and of its social context – variables which play no part in the life of a quark or a carbon atom.

These two factors drastically limit the use of the atomizing approach on social phenomena. That is why appeals to 'holism' to balance reductionism in this area are absolutely necessary. Such appeals are, incidentally, often found to be needed in physics, chemistry and mathematics too for very similar reasons. There too the role of parts simply cannot be described without reference to a wider context. But that is not our present business.

TRYING TO SIMPLIFY MOTIVATION

Granted the difficulties of atomizing conduct, theorists have naturally tried other forms of psychological reduction besides the analysis of whole to parts. These, too, have often been seen as modelled on the physical sciences. As we noted in Chapter 3, p. 39, attempts to understand motives often proceed, not by atomizing them, but by lumping many of them together under one main heading. The chosen motive tends to be one which is viewed as deeper and cruder than those it is called in to explain, for instance sex or pleasure or self-interest.

This approach to psychology has usually been seen, since the Renaissance, as a 'scientific' one, producing a more systematic account than existing ideas on motivation and therefore worthy to demote them to the limbo of mere 'folk-psychology'. The chief scientific model for this move may well have been Galileo's extension of earthly physics to cover the heavens, replacing the two distinct sets of laws envisaged by medieval thought. And of course there have been many other cases in science where theories proposed for limited areas have been successfully widened.

Whatever the model, reducers in the Hobbesian tradition strengthened this impression of a base in science by linking their theories of motive with the reduction of mind to body. As Hobbes put it, 'life itself is nothing but motion'.[4] This, however, is an independent metaphysical move of a quite different kind. It does not support sweeping simplifications of motive and is not supported by them. Those simplifications have to justify themselves on their own merits, by proving useful in psychology itself.

They would, however, have to be quite extraordinarily useful if they were really to justify relegating all other thought on the subject to the extent required. And as it happens, the scientific status of psychological reductions like those of Freud has itself come under sharp attack from theorists anxious to narrow the frontiers of science. The ideal of objectivity has increasingly been held to demand a freedom from bias which views like these cannot easily claim – more especially since there are many of them and they often contradict one another. Accordingly, attention has shifted away from these direct analyses of motive towards the wider, more abstract reduction of mind to body.

WHAT IS OBJECTIVITY?

This reduction is often seen as bias-free. It is welcomed because 'science' as such is seen as something that corrects bias. Scientific thought is conceived, not only as containing no factual assumptions from outside itself, but also as stemming from no motives other than pure abstract curiosity. Such thought is deemed to have no moral or political aims. It is therefore welcomed as a way of correcting the various biases that are obviously liable to affect the social sciences and humanities. These studies naturally do often attract people who want, not just to understand society, but also to act on it, to produce reforms and revolutions. They may want, for instance, to increase freedom, to cure injustice, or of course to do something less respectable.

These clashing biases are indeed both dangerous and confusing and, as our society grows more complex, more and more of them proliferate. Alarm about this seems to have played a large part in the campaign to bring these studies more closely into line with physical science. As we have seen, for this end, two distinct strategies have always been used. One affects method, treating the social sciences as part of 'science' and urging them to become more impartial by using methods closer to those of their physical brothers. The other is metaphysical, proclaiming that in any case only material entities are real, so that social and personal life, however studied, will always remain in some sense provisional and illusory.

Both proposals are still in vigorous use. But the first one, which is the more easily understood and tested, has repeatedly led to disillusionment. The attempt to impose an entirely impartial approach on those sciences which are concerned with matters of direct human concern does not make much sense. Indeed, it is not easy to apply even within physical science itself. People often do not notice what an extreme demand it is that science should confine itself to reporting and collating unmistakable facts, without committing itself to any special ways of interpreting them. This demand, if it were complied with, might well paralyse scientific enquiry altogether. As we have noticed, interpretative schemes are needed for selecting one's facts in the first place, and they always do express a more general outlook on life.

Certainly the demand would make it impossible for any scientific doctrine to have moral or ideological consequences. For this reason, if for no other, it has been made only half-heartedly. People who otherwise might support it are usually still anxious that the world at large should treat scientific ideas as important. You cannot eat the cake of an exclusively pure status and expect also to keep any influence on practical affairs.

This limitation on impartiality does not, of course, mean that we have to endorse the wilder relativistic claims of some sociologists and historians of science. We do not have to say that theories are only arbitrary, expressive cultural constructions. But theories do come out of cultures and have to draw their language from them. The ideal of objectivity, like other ideals, is a distant aim, a beacon among the other beacons that guide us, not a place that we ever reach. It has no general dominance over other ideals. Scientific work ought not always to be objective, if that means being impartial on serious moral questions. Moral considerations often do enter into the directing of scientific enquiries – for instance in medicine – and clearly it is right that they should do so. They ought, for example, to have prevented Nazi research.

THE CASE OF EUGENICS

It is important to see how this kind of influence can work even among scientists who suppose themselves to be conducting a straightforward impartial enquiry. For instance, it is remarkable to see how the journal *Nature*, which was then as now the respected mouthpiece of British science, expressed throughout the earlier years of this century strong support for 'negative eugenics' – that is, for widespread sterilization, not as a control on population numbers, but selectively, to restrict certain groups. In 1924 it launched a vigorous editorial campaign to that effect. Intelligence testers (its editorial explained) had already discovered

> that 'a large proportion of the slum populations consists of ... "morons" – that is, of mental defectives of comparatively high grade. These people are lacking not only in intelligence but also in self-control, which is the basis of morality' ... The journal even outdid the Eugenics Society by calling, not merely for 'voluntary' sterilization of hereditary mental and physical defectives, but for compulsory sterilization as 'a punishment for the economic sin of producing more children than the parents can support'.[5]

The crucial point here is that this campaign was not intended to be an incursion into politics. The journal strongly disapproved of such incursions. It supported a general ideal of scientific purity, reproving Marxists like J.D. Bernal who linked science with politics. But eugenic data of the kind that it relied on did not strike its editors as political material. Those editors, like very many ordinary scientists of the time, saw them as simply something internal to science, something as solid as the law of gravitation, truths so uncontroversial that there could be no doubt about the practical lessons to be drawn from them. 'Humanitarian sentiment acting in ignorance of *the laws of biology* is a most dangerous thing and produces devastating results' (emphasis mine).[6] Thus, as late as 1936, *Nature*'s leading article still treated the matter as scientifically straightforward:

> Dock labourers and miners figure prominently in the overproduction of children, and it is worthy of note that in both groups there is a large proportion of the Iberian element in our population from Wales and Ireland ... *It is the reproduction of this class that we wish to prevent.*[7] (Emphasis mine and Werskey's)

Soon after this date *Nature* did change its policy, but interestingly, this was not because conflicting factual data had forced a change in scientific judgement. No controversy intervened. The disturbing factor seems to have been simply the embarrassing likeness between this line and Nazi propaganda. That caused scientists to notice the matter. Once they did so, brief reflection was enough to make them expel from science an emperor who had never had any clothes. The cultural factors that had determined their biased selection of data had shifted. What this shows is how important such cultural factors are, even in decisions about what is to count as science, and how essential it is that scientists should be aware of them. The remedy is not to isolate science antiseptically from the rest of culture, since that is impossible. It is to understand its relation to the rest of thought, its position as an institution within that culture.

6

CONVERGENT EXPLANATIONS
AND THEIR USES

— ·◆· —

EXAMPLE – THE CASE OF
EVAN JONES

IT IS time now to describe more fully how a non-reductive, pluralistic kind of explanation can work.

Enlightenment thought has concentrated so strongly on a simplified, linear pattern that there seems to be a real difficulty today in grasping any alternative. How can we say, without mere confusion and anarchy, that there can be many different ways of describing something, all of which – except for a few charlatans – are legitimate in their own terms, can be usefully related, and do not need to be reduced to one another? Our current learned language makes this sound mysterious. But it is something that we do every day.

When I touched on this matter earlier, I used the image of a mountain which is looked at from many sides. I pointed out that, since we can travel between these sides, there need be no difficulty in building up a reasonably unified composite picture of this mountain. Certainly the image implies that we shall never have absolutely complete knowledge of it, but then that is surely obvious in any case. Later, I partly cashed this image by speaking of a tree which is described and understood in different ways by people with various special ways of understanding it. But we need now to speak more directly and literally still. We had better take a particular concrete case.

The one I propose is simply the question why a particular person – Evan Jones – is not dead yet? He is one of those people who defy the doctors. He ought to have been dead two years ago. But he is still, not just alive, but a power in village affairs. Medically speaking, he is a case of 'spontaneous remission', statistically unusual, but not numerous enough to affect the received doctrines. Informally speaking, he's a case of free will. He's got a great deal to do, and he isn't prepared to go until he's done it.

As he sits at his kitchen table, vigorously drafting a letter to the local council, we can imagine the converging, but not necessarily competing, forms of explanation as raying out from him on all sides. On his right, there are the physical sciences. The medical explanations that they offer relate to pneumoconiosis, complicated by all the oddities of his personal physique and family history. The genetic ones illuminate that family history. The geographical ones deal with the local climate and conditions, particularly the mines.

Further out than these and other applied sciences lie the pure sciences which order them and on which they draw – various branches of biology, chemistry and physics, in that order. Further out still lies mathematics. No doubt these last sciences, so ordered, are in some sense 'fundamental' for their own area. But they have no direct relevance to explaining Evan's situation. They can of course bear on it indirectly, by affecting the intermediate applied sciences, such as medicine. A new chemical discovery might affect his diagnosis or treatment. But this would obviously be a fairly minor contribution to his case. It could not be a ground for saying that any of these sciences supplied the real 'fundamental explanation' of it.

On Evan's left lie the historical and political explanations. These include facts about obvious things like labour and housing conditions, wars and national politics, and also more local cultural matters, such as choirs, schools, chapels, libraries and debating societies, which contribute to the meaning of his life. The social sciences, which lie near to these, may cast useful light on these various institutions. And the institutions are important to him. But in an obvious sense he is not very important to them, because they are corporate and he is only one individual. He does not necessarily represent the average at all.

All these enquiries can be used to explain a great deal about Evan's situation, but not much about why his response to it differs from other people's. In order to reach this angle, we shall need to *know* him, in the ordinary but very significant personal sense, as well as knowing general facts about him. (It is quite extraordinary that epistemology has wholly neglected this very important aspect of knowledge – knowing people.) We shall have to see him, not as a statistic, but as an individual, taking in his own point of view and that of those nearest to him.

Here we find a range of quite different, more private and subjective conceptual schemes which ray out, as it were, behind him. They are those used by himself, his family and others who do know him personally. These schemes are ancient, widespread and in a sense informal. That, however, does not at all mean that they are crude. As anthropologists find when they try to articulate the systems of other cultures, they can be extremely complex. A great deal of sophisticated thinking goes on in this area, building up the thing called wisdom. This is the fertile seed-bed out of which all more formal and theoretical ideas have arisen. It is their country of origin and it furnishes the climate without which they cannot survive.

Here live an enormous battery of practical notions about how to live and how to act – notions which have meaning only at this subjective viewpoint – and also other ideas that Evan uses to think about his own hopes and purposes. Many of these ideas are shared also by people who think themselves much more sophisticated than he is. They go to shape the beliefs about ethics, aesthetics and religion by which he articulates those purposes. Beyond these beliefs, and ordering them in a way somewhat like that in which the pure sciences order the applied ones, lie a number of more public and objective thought-systems which he has taken from his culture, systems ranging from traditional views about nationality and politics to the latest theories about psychology.

THE DISTINCTNESS OF SUBJECTS

Something very important arises here. We need to notice briefly a point which will become central later, namely, just how different the subjective viewpoint is from all the others so far mentioned. It is in fact the only one where there is literally a *viewpoint* – a point that views, a unit of experience, a subject that actually makes its own observations. When we talk of other viewpoints, such as the medical or the economic, we are thinking of them as positions that a subject can move to and use. They are ways of thinking that he can decide to employ. But the subject is the one who does this deciding.

This discontinuity between a subject and all its objects is crucial for our whole argument. It is so wide a gap that it has always defeated systematizers. The typical twentieth-century response to it has been to evade it by ignoring subjects entirely, as if attending to them were 'subjective' – that is, biased. By contrast, earlier thinkers commonly separated out subjects as distinct things or substances, souls or minds.

The trouble with this approach has been that these 'substances' are easily conceived on the model of ordinary physical things like chairs and cabbages. They then seem like strange, quasi-physical, ghostly objects added to the world's furniture. The earlier philosophers did not actually have this crude notion of substance. They used the word more widely for anything that there is.[1] But even with this wider notion, there was always a real difficulty about seeing how the word could bring together such very different items as minds and bodies. Theorists longed to find some stronger way of unifying the points of view.

Many philosophers, ranging from sceptics like Hume to castle-builders like Hegel, tried the opposite reduction. They evaporated matter away by reducing it to mind. All substances are then seen as built up out of experience and consisting of perceptions or of soul-stuff. The systems that result from this move are probably a little more intelligible than those that show the world as made only of matter, but that is not saying much. Neither kind of reduction works. Both, after their first moves, encounter a mass of difficulties. A world without objects is not much easier to conceive of than a world without subjects. Only an arbitrary and unstable unity can be imposed by making one aspect swallow up the other. The original dichotomy still remains.

The important question here is not 'what stuff is the world made of?' That is the question from which the pre-Socratic philosophers started and it has to lead finally to modern physics, though its propounders tried stuffs such as water and fire and spirit and little hard atoms and many other things on their way there. But the question that we much more deeply need to ask is the wider one about the relation between the subjective and objective viewpoints. Thomas Nagel puts forward this question at the outset of his book *The View from Nowhere*. He says,

> This book is about a single problem: how to combine the perspective of a particular person inside the world with an objective view of that same world, the person and his viewpoint included ... It is the most fundamental issue about morality, knowledge, freedom, the self, and the relation of mind to the physical world.[2]

Nagel does not expect any complete enclosing system to emerge. But he proposes to make

> a deliberate effort to juxtapose the internal and external or subjective and objective views at *full strength*, in order to

achieve unification where it is possible and to recognise clearly when it is not. Instead of a unified world-view, we get the interplay of these two uneasily related types of conception, and the essentially uncompletable effort to reconcile them.[3] (Emphasis mine)

This proposal is revolutionary because of its emphasis on taking both angles seriously, 'at full strength'. Earlier unifiers have always belittled one of these aspects so as to leave room for the other. It has been assumed that ultimately we must come off the fence. We have to choose between them, and in recent times it has been taken for granted that the objective one must be the winner. Thus Daniel Dennett thinks it necessary to begin his study *The Intentional Stance* with 'a tactical choice' and accordingly writes,

> I declare my starting-point to be the objective, materialistic, third-person world of the physical sciences. This is the orthodox choice today in the English-speaking world.[4]

So it is, but that doesn't mean that it makes sense. Nobody has the objective, third-person world as their only starting-point. As Nagel points out, objectivity cannot possibly be the only ideal that guides thought. Objectivity is only one technique used in understanding;

> To acquire a more objective understanding of some aspect of life or the world, we step back from our initial view of it and form a new conception which has that view and its relation to the world as its object. In other words, we place ourselves in the world that is to be understood ... The process can be repeated, yielding a still more objective conception.

This detachment is, as he says, often helpful, but there are sharp limits to its usefulness:

> Although there is a connexion between objectivity and reality – only the supposition that we and our appearances are part of a larger reality makes it reasonable to seek understanding by stepping back from the appearances in this way – still *not all reality is better understood the more objectively it is viewed.* Appearance and perspective are essential parts of what there is, and in some respects they are better understood from a less detached standpoint.

In fact, since the backwards step can be repeated,

> the distinction between *more subjective and more objective views is really a matter of degree*, and it covers a wide spectrum . . . The standpoint of morality is more objective than that of private life, but less objective than the standpoint of physics. (Emphases mine)

Stepping back to a more objective viewpoint is rather like using a more inclusive map with a smaller scale. The map of the British Isles is the right tool for envisaging the whole country, but it is not much help for finding your way around Shepherd's Bush. On many questions, including detailed factual questions about experience as well as emotional and practical matters, the subjective point of view must be directly consulted, and on quite a lot of them it is the more important. Above all, it is the point of view of the agent. For practical questions this means that, if it is not included and given its right weight along with the more objective angles that get considered, nothing will actually get done. And that, indeed, is what very often happens.

FUNDAMENTAL FOR WHAT?

We will have to say more later about this distinctiveness of subjects, which is a vital point for the understanding of freedom. For the moment, however, we are still occupied with the question about the relation between different kinds of explanation. And the question here is: ought any of these conceptual schemes to be picked out as giving the *real* explanation of Evan's situation?

It is surely most obscure what would be the point or force of doing this. Different kinds of questions need different answers. When we dismiss one explanation as superficial and welcome another as deeper, they need to be answers to the same question. With a very general question – 'why isn't he dead yet?' – the real answer is simply the one which best resolves that particular difficulty.

Answers which don't do this – such, for instance, as the contributions of chemistry and physics to this particular enquiry – are of course not illusory. They are the right answers to different questions, but for this enquiry they are unhelpful and irrelevant. In this context, then, they are in no sense fundamental. And in fact we do not always need to pick out one answer as the real or fundamental one at all. Very often we can accept several partial answers as contributing.

Sometimes, of course, there is a special reason for deciding which of these is the most important, and this is what brings in talk of 'reality'. We may indeed say that we have not found the real explanation yet. But this means we have not yet got one that fits the question properly. *Real explanations, like real coffee or real cream, are ones that meet suitable standards and succeed in doing what is asked of them.* They are ones that answer the right question. There is no metaphysical short cut to finding them by proving certain kinds of entity – such as electrons or brain cells – to be more real than others, such as thoughts or feelings.

THE DREAM OF MONOLITHIC ORDER

The idea of a single, ruling scheme, underlying all others, answering all questions and transcending all difference of viewpoint, has fascinated thinkers since the dawn of philosophy. That idea reflects a general unifying aim which is indeed a genuine demand of reason. Unifying a field of enquiry certainly does explain it – but only in so far as the process does not distort it. Beyond a certain point, gains in comprehensiveness seem always to be paid for by Procrustean damage in some areas. And the reason for adopting one scheme rather than another always reflects a bias in the first place.

Only while we protect ourselves by abstraction can we forget this kind of bias and find simplifications plausible. Whenever we keep an eye on reality, endless kinds of question arise, and refuse to be stereotyped in the interests of tidiness. Only for bad reasons do we confine ourselves to a single viewpoint, or back one thought-system to crush all the rest. The case that we are considering here – which is by no means a specially complicated one – makes this very plain.

In real life, nobody dreams of privileging a single standpoint. All those concerned, including Mr Jones himself and including his doctor, constantly check their own special angle by looking at a number of others. The whole person is plainly so complex that a single account cannot be adequate. Any one account can fail us unexpectedly, as medicine has done here, and simply have nothing more to say. But it is a generally known fact about medicine that it can so fail, and we know roughly what measures to take when it does.

In these emergencies we do *not*, unless from some special indications, move at once outward, to the more abstract sciences that lie beyond medicine. Mostly, they will not help us at all. They do not have answers to the questions we need to ask. Instead, we move round to certain indicated positions on the other sides, wondering whether what we can see from there will complete the pattern. Both the historical and the subjective angle may well be obvious ones to consult.

If these give us good, convincing results, it will be perfectly sensible to say, for instance, that 'the fundamental factor is his belief in socialism' or 'his position in the village'. Saying this does not clash at all with the proposition that physics and chemistry rank as more 'fundamental' sciences than medicine, because the direction from which they count as fundamental is a quite different one. And again, if the first explanations that we meet when we look from these various angles seem superficial, it will be quite proper to look for a 'more fundamental' one in that same area.

7

TROUBLES OF THE LINEAR PATTERN

FUNDAMENTAL FOR WHAT?

CERTAINLY WE expect consistency between different viewpoints. Since we know that the subject is one person, to accept contradictions about him or her would be a damaging piece of scepticism. But since we also know that this subject is incredibly complicated, mere temporary, surface inconsistency does not surprise us or alarm us much. We put it down to faults in our conceptual schemes. It calls for hard work to improve them, but not for despair. And it certainly does not call for the drastic remedy of making one thought-scheme a dictator, dubbing it fundamental and giving it a general licence to override all others.

No doubt a single universal pattern would be more intellectually satisfying, if we could only find one that really worked. But patterns stop being satisfying when they do not fit the facts. The discontinuities between the viewpoints are real, at least for non-omniscient beings like ourselves. Mr Jones is solid, opaque, three-dimensional like the mountain. We are never going to know all about him. It is to bring this out that I have set him there, solidly at his kitchen table. The picture of different enquiries raying out from him in different directions is designed to correct the idea that they are piled up in layers with the best one at the bottom, so that we need only dig down through the rest to find the final truth. That is the idea so misleadingly suggested by the metaphor 'fundamental'.

To grasp the contrast, it may be worth while to look briefly at this traditional linear up-and-down arrangement. It is well set out by Edward O. Wilson in his book *On Human Nature*. Wilson invents the word 'anti-discipline' for the study which lies next below any given enquiry in the reductive pile. He then explains that

> Biology stands today as the anti-discipline of the social sciences. By the word 'anti-discipline' I wish to emphasise the special adversary relation that often exists when fields of study at adjacent levels of organization begin to interact.[1]

This relation, he says, often leads to a dialectical conflict ending in a synthesis that alters both, and he gives examples of this process from various natural sciences. This sounds as if the lop-sidedness which we noted earlier as a feature of reduction is being corrected.[2] Wilson seems to show both studies as standing on equal terms, conducting a two-way exchange, not a hierarchical conquest. To understand how this will work, we ask what kind of synthesis is now to be expected.

Well, (Wilson replies) biology is now to contribute 'the conceptual foundation of the social sciences'. That is, it will shift their scope altogether by indicating the proper topic for their attention. There is no suggestion that these sciences will return the compliment by shifting biology in its turn to a new subject-matter. But the social sciences must be moved because 'the core of social theory . . . is the deep structure of human nature, *an essentially biological phenomenon* that is also the primary focus of the humanities' (emphasis mine). Indeed, 'the scientific materialism embodied in biology will, through a re-examination of the mind and the foundations of social behaviour, serve as a kind of anti-discipline to the humanities'[3] as well.

NEUROBIOLOGIZING ETHICS

In particular, Wilson says, it must reshape ethics, which is also to have its focus moved to a wholly different subject-matter, namely, neurobiology. As Wilson put it in an earlier book, 'The time has come for ethics to be removed from the hands of the philosophers and biologicised.'[4] He explains that this will, in fact, be the first time that anything but thoroughly amateur thinking has been done on the subject at all:

> Like everyone else, philosophers measure their personal emotional responses to various alternatives as though consulting a hidden oracle.

> That oracle resides in the deep emotional centers of the brain, most probably within the limbic system, a complex array

of neurons and hormone-secreting cells just beneath the 'thinking' portion of the cerebral cortex . . . The only way forward is to study human nature as part of the natural sciences . . . Neurobiology cannot be learnt at the feet of a guru. The consequences of genetic history cannot be chosen by legislatures. Above all, for our own physical well-being if nothing else, *ethical philosophy must not be left in the hands of the merely wise* . . . Only hard-won empirical knowledge of our biological nature will allow us to *make optimum choices among the competing criteria of progress.*[5] (Emphases mine)

Wilson, like Skinner, offers a quite new hope of solving the world's practical problems by discovering a new set of facts, though his facts are different ones. Like Skinner too, he seems not to know that his offer is meaningless because ethical problems are not themselves factual ones at all, but are practical problems about what to do and how to think. 'Optimum choices among the competing criteria of progress' may indeed need all sorts of good factual data. But the point where they concern ethics is that they are choices among standards, among values, among ideals. Moral thinking involves using and developing the practical conceptual schemes which have been evolved to sort out these choices. And we have noticed that these schemes are complex. Working on them is not much like consulting an oracle. There is nothing very mere about wisdom.

Manifestos like Wilson's cannot yield a tribal victory. They call for a split verdict. Facts about human nature are indeed important among the many ranges of facts that are needed for moral choices. What we know about human nature does tell us facts, often crucial ones, about the range of possibilities open to us. And we often need these for making a choice. Here Wilson is right, as against the anti-naturalist theories that we shall be considering later. But he is plainly wrong, first, in supposing neurobiology to be the only source for these important facts. We find out about human nature from a thousand sources, most obviously from everyday life and from history. Without those other sources brain science would not have the concepts and assumptions from which its investigations start. Moreover, it must continually use these outside concepts and assumptions to check the meaning of its work – for instance, in asking experimental subjects to report on their experiences. The reason for giving neurobiology this startling priority evidently lies in the linear image of how the sciences are arranged – the pattern of successive levels of depth – which Wilson is still taking for granted.

WIDER YET AND WIDER . . .

This same image, however, surely also accounts for something more serious. It explains Wilson's exaggerated hopes, his expecting from neurobiology a kind of work that it could not possibly do. He takes it that, because we need our limbic systems in order to think about standards, therefore we shall find the right standards by investigating the limbic system. What ethics would amount to on this plan is quite obscure. Perhaps we would then say, for instance, 'Since x and y have now been found to occur in the hypothalamus, we can safely conclude that art matters more than science', or 'that we will be justified in invading Ruritania'? This is much like suggesting that the way to advance mathematics or logic is to explore those parts of the brain that mathematicians and logicians use – a proposal which nobody seems yet to have made. As far as I know, neither Wilson nor anybody else proposing this kind of reduction gives any illustration of the way in which it could produce moral thinking, any more than Skinner did. Yet neurobiology has surely been around long enough to provide examples, if the matter has been thought through.

Perhaps the suggestion really centres, as Wilson's mention of 'materialism' suggests, on getting rid of the soul. Perhaps the mere removal of religion is expected to make all moral problems vanish? I suspect, however, that the confidence with which this kind of suggestion is put forward does indeed rest mainly on the powerful, unquestioned pattern of superimposed levels itself. That pattern carries with it a certainty that the 'deeper' study is always the more authoritative. And the direction of depth is never in doubt. Physics still remains where Descartes put it, at the bottom of the pile. The series – which makes quite good sense as far up as molecular biology – is casually prolonged to cover all other studies.

All this has a cheering result for academic imperialists. It means that no enquiry need really take the shallower, more amateur studies above it seriously. Conciliatory talk about creative dialectic and mutual consultation may serve to lower the temperature, but at the level of real life the hierarchy will still prevail. The proper behaviour-pattern is to conquer these other studies and then to devour them. As Wilson put it in his less guarded days in *Sociobiology*:

It may not be too much to say that sociology and the other social sciences, as well as the humanities, are the last branches of biology waiting to be included in the Modern Synthesis.

One of the functions of Sociobiology, then, is to reformulate the foundations of the social sciences in a way that draws these subjects into the Modern Synthesis ... Having cannibalized psychology, the new neurobiology will yield an enduring set of first principles for sociology.[6]

This imperialism is backed by a startlingly naïve, lyrical faith in the reductive prospects to be expected from this conquest:

When man has achieved an ecologically steady state, probably by the end of the twenty-first century, the internalization of social evolution will be nearly complete. About this time biology should be at its peak, with the social sciences maturing rapidly ... The transition from purely phenomenological to fundamental theory in sociology must await a full, neuronal explanation of the human brain. Only when the machinery can be torn down on paper at the level of the cell and put together again will the properties of emotion and ethical judgment come clear ... Stress will be evaluated in terms of the neurophysiological perturbations and their relaxation times. Cognition will be translated into circuitry. Learning and creativeness will be defined as the alteration of specific portions of the cognitive machinery regulated by input from the emotive centers. Having cannibalized psychology, the new neurobiology will yield an enduring set of first principles for sociology ... Skinner's dream of a culture predesigned for happiness will surely have to wait for the new neurobiology. A genetically accurate and hence completely fair system of ethics must also wait.[7]

I have discussed these dreams in more detail elsewhere and I need say no more about them now.[8] The point for our present purpose is the way they illustrate the hierarchical nature of the linear reductive series. The upper levels are not seen as making any real contribution to the profounder ones below. There is no mutually useful dialectic. The Modern Synthesis that Wilson has in mind is simply the union of Darwin and Mendel, supplemented by more recent evolutionary theory. His more placatory talk in *On Human Nature* of a mutually beneficial dialogue with non-scientific studies does not signal any real intention to listen to them. Cannibalism is, after all, a one-way process.

Wilson is instructive because he is unusually uninhibited about expressing his confidence in such plans, and in particular about his imperialist ambitions. But the views he expresses are not unusual at all. Like Skinner, he is important not because he is an original thinker but precisely because he is not. He is a lightning-conductor that picks up whatever is in the air and gives it inflammatory expression.

IS THERE A CHOICE OF DESPOTS?

If this objection to intellectual despotism is fair, it is worth asking whether any particular despot is worse than any other. Are there better and worse choices for supposedly ultimate and 'fundamental' explanations?

In a way, they are all equally wrong. It is just as mistaken to be exclusively religious, or exclusively subjective, or exclusively historical, as it is to be exclusively physical, though it is not so fashionable, and that may make it look milder. It is interesting that the word 'reductive' is largely used for reductions to the physical sciences. No doubt this is largely because these are seen as reductions of wholes to parts, and eventually to ultimate particles. Other exclusive approaches, however, can be bad in a very similar way. For instance, a colleague might see Evan Jones only in rather crude Marxist terms, as a victim, a revolutionary, a reformist, or perhaps a blackleg and class traitor. Or again, in crude Freudian terms, as a case of arrested development, or in crude religious terms, as only a soul to be saved in some specialized manner.

This psychological and political reduction has been quite as common as the physicalist kind, and it has done at least as much harm. Ought all these approaches too to be called reductive, or are they just ordinary cases of being narrow-minded, obsessive, ignorant, arbitrary and inhuman? Probably the answer is that the word reductive is now so closely linked with the project of extending the physical sciences that it must keep that connection. But this one-sided interpretation is perhaps unfortunate. Narrow-mindedness that expresses itself by restriction to any single conceptual scheme is a phenomenon of some interest that could do with a name of its own.

Is there some single worst form of it? At an obvious level, the most dangerous forms do seem to be the ones which are most dramatizable, most colourful, most visibly relevant to life – the ones with the most imaginative force. But this kind of persuasiveness naturally belongs to different kinds of simplification in different epochs. The soil in which myths can grow constantly changes as background ideas alter.

For instance, in the days when witch-hunting and religious perse-cution were rife, an exclusive concentration on religion might well be the most dangerous choice. This history fully explains the anti-theism of Voltaire and Hume, and it underlies the simple-minded, atheistical converting zeal of those who still follow them. Out in the world, however, things have changed. Christianity has not now much direct political clout. Instead, the West's potent myths today all in some way express the exaltation of machines.

MACHINE HYPE

Beyond the mere obsession with, and confidence in, actual machines, which can itself get surprisingly near to worship, they tend to take the form of belief in vast, mechanical super-processes, within which we are tiny cogs. By a disastrous equivocation, these processes are seen now as irresistible, now as all-justifying. Acceptance of them is profoundly fatalistic. Their message that effective action is impossible can actually paralyse people.

This danger was very plain in Marxism and it is surely no less so in the exaltation of market forces. These systems have indeed a great deal in common, which no doubt explains how easily people convert from one to the other. By contrast, the merely physicalist reduction of minds to bodies and eventually to physical particles may seem to be purely theoretical and so relatively harmless. And if it had appeared alone it might possibly have been so. If we are told that we are – really – just arrays of electrons, this news may seem like something totally detached from all our real concerns.

Does it have any practical consequences at all? Ought we, for instance, to stop feeling free? If so, what are we supposed to do about it? Should we ring up the nearest physics lab and ask them what we are going to do next? They won't know, and nothing seems to follow. If, on the other hand, we had decided to view ourselves as only victims, or only souls to be saved, or only cogs in an economic process, our lives could be radically changed. Of course there might be reasons for making that change. But a mere obsession with a single despotic conceptual scheme doesn't seem like the right sort of reason at all.

HOW HARMLESS IS PHYSICALISM?

Physicalist reduction on its own can, then, seem to have no bearing on practical life. And those academics who treat it as a purely formal intellectual system do indeed often see it as isolated in this way from the rest of thought. In the general intellectual history, however, it has not been thus isolated. From Hobbes on – indeed, from the Greek Atomists – powerful propagandists have used it as a tool to promote a wide variety of colourful world-views with strong moral consequences.

This has been possible because the physical reduction has not been proclaimed on its own as a piece of pure theory but as the end of a route lying through biology and psychology, where matters of immediate concern arise. When minds have been reduced to bodies, bodies have also been alleged to be ruled by certain overwhelming particular patterns of motive, such as egoism or sexuality, or to be helpless in the grip of wider historical processes which made certain particular kinds of action advisable and others pointless. Moral controversy has been the habitat in which this reductive thinking has been developed, and it is still extremely active there.

It would surely be hard for anyone endowed with normal sensitivity to current moral issues to carry out the academics' project of thinking about formal reduction without reacting at all to this colourful context. And many distinguished reducers have said plainly that ideological factors have indeed determined their intellectual course. Thus Francis Crick writes in his memoir:

> This loss of faith in Christian religion and my growing attachment to science have played *a dominant part* in my scientific career . . . I realised early on that it is detailed scientific knowledge which makes certain religious beliefs untenable . . . *What would be more important than to find our true place in the universe by removing one by one these unfortunate vestiges of earlier beliefs?* . . . It seemed to me of the first importance to identify these unexplained areas of knowledge and to work toward their scientific understanding.[9]

Wilson has repeatedly declared a similar campaigning purpose,[10] and so has Richard Dawkins, for instance in a recent article entitled 'Is God a computer virus?'[11]

DO TOADS BENEATH THE HARROW LOSE THEIR STATUS?

In this wider moral context, physicalist reduction itself may seem to play a minor part. But it does have one specially alarming feature.

It explicitly cuts out the subjective angle. It reduces people, not just to their bodies, but to something which they do not feel themselves to be at all and may not even have heard of – quarks and the like. Though it may be obscure just what follows from this, it does seem natural to conclude that the topic must now be handed over entirely to the experts. What little authority we once thought we had for speaking about our own lives has vanished.

That authority has, of course, always been limited. We have always known that we are very far from knowing all about our own lives. Experience teaches us that other people round us can often put us right – for instance, by correcting our memory, or by pointing out that our motives are not what we think they are. And we accept, too, that the lore of our own culture, and that of other cultures, can often correct the local verdict. But still, on certain matters directly concerning our own personal lives it did seem that we and those around us knew what we were talking about. It did seem that, there, we were the ones who should be asked first.

At the point where all the concepts that we normally use get dismissed as 'folk-psychology', this stops being true. At that point, the experts take over. We are placed in something like the alarming situation of people taken to a mental hospital where the staff ignore everything that they say. And if these experts are no longer even medical ones but experts in physical science, we have lost even that continuity between their pronouncements and our own point of view that medicine officially offers. Having lost the traditional psychology of motive that even Freudian and Nietzschean theorists still acknowledged, we have no longer a shared, common language. Those theorists may have been overbearing, but in principle they did appeal to the same standards as the rest of us. What they said would only be believed if at some point it was borne out by general human experience. If, however, physical science has the last word, general human experience becomes as irrelevant as it would be in determining the exact position of the sun.

8

FATALISM
AND PREDICTABILITY
·◆·

PREDICTION IS NOT
THE REAL DANGER

THIS ELEVATION of an academic sect to a position of authority over everyone else, not just on particular facts but on the whole of people's lives, seems to be the central threat that reduction poses to our ideas of freedom. The objectionable point does not lie in admitting that we do not know everything about ourselves. It lies in naming a particular branch of learning, and therefore a particular set of other people, as being the ones who do know about us. It lies in treating all the findings of human experience, however carefully arrived at, as empty unless practitioners of that study approve them, thereby handing these people unexampled authority and power.

These pretensions could never be justified because there just is no court, independent of human experience, which could validate their pronouncements on that experience and show reason to believe them. The experts would be just as dependent as Freudian enquiry has always been on producing results coherent, at some stage, with the ways in which experience works. Since experience is always the experience of people who act, and since these people need to order their thoughts by the concepts adapted to make action possible, observers who try to avoid attending to those concepts cannot hope to discover anything useful.

What these pretensions can do, however, is to undermine the confidence which people rightly have in their own power of judging their actions. Reductive claims can promote particular views on personal identity – on what a person actually is – which carry whole fatalistic ideologies along with them, and they can back these views by the authority of science. This, I believe, is the point at which reductionism really threatens freedom.

The central trouble is not that reduction makes conduct look predictable. Predictableness in itself is not necessarily a threat and its relevance to this matter has been greatly exaggerated. In many ways, after all, we want and try to be predictable. When heroic people who have been caught by the secret police make just the determined resistance that their colleagues expect of them, nobody thinks they have been proved to be mere machines. Trustworthiness, reliability, responsibility are all virtues. More widely, too, a world in which our actions were completely unpredictable would certainly not be one that would restore our confidence. If, like a roulette wheel, we might do just anything, we could hardly be said to be acting at all. We expect and demand certain kinds of regularity, both in our own conduct and in those of others.

This demand is not just a regrettable bourgeois weakness. Our natural temperaments absolutely require a certain amount of regularity, just as they also require a certain amount of change. Certainly we can sometimes feel a kind of restlessness that seems only to want change, a mood that wants surprise and cannot bear repetition. We can get bored with having snowdrops every January and sigh for orchids or explosions instead. This mood can lead us at times to value changed conduct for its own sake, and we sometimes make trouble in order to get it. But it will only be worth our while to do this if we can predict that the change we want will follow. If it does not, the failure of our prediction will disappoint us and our search for the unpredictable will be unpredictably frustrated.

SPONTANEITY AND COMPULSION

What freedom demands is not, I think, this mysterious negative property of unpredictability at all. It is something more positive and harder to define called spontaneity. What matters is not whether our acts can be predicted, but whether they are our own, whether they come from the heart and are what we mean. Now it is certainly true that words and actions which come from the heart cannot usually be predicted in any detail by other human beings, though perhaps an omniscient god might predict them. And it is true that unfriendly human beings who *do* predict our actions can gain a dangerous power over us. But when we complain of something's being 'predictable', we are objecting to something else besides that power.

The kind of case where predictability is really alarming is, I think, that where an action seems to be enforced, whether by suggestion from outside or from parts of the self with which, for one reason or another, we do not identify. Agency has then been lost. In cases of rigid habit or overwhelming passion, the act seems compulsive and bystanders can do the predicting, just as they could in cases of threats or strong suggestion from outside. Occasionally we may even predict it ourselves, but to do so is to abdicate our authority. If we do that, we have stopped thinking of ourselves as acting. The uniting pattern of the whole has been broken. The agent as a whole has ceased to own the action which has been assimilated instead to surrounding processes.

By contrast, prediction that proceeds through considerations internal to the pattern of our own life, considerations that we ourselves recognize, is not an offence, indeed it may even be demanded. People sometimes say 'You ought to have known that I would never do that.' The core point, then, really concerns personal identity. We are free – not if we do something unpredictable, but – if our act is our own. This ownness is certainly a puzzling notion. We shall be repeatedly concerned with it. The difficulty of reconciling it with a true recognition of our slight and misty nature – our fragmentedness, our deep dependence on the biosphere and the many doubts about how far our consciousness extends – is a central topic for this book. But since human life cannot be conducted without this sense of ownness, it is sheer humbug to dismiss it as a mere folk-psychological illusion.

STRAWSON ON RESENTMENT

Peter Strawson has made this point well in his seminal article 'Freedom and resentment'.[1] He notes how sharply current thought tends to divide into two stereotyped streams about acceptance of determinism. As he says,

> Some – the pessimists perhaps – hold that if the thesis [of determinism] is true, then the concepts of moral obligation and responsibility really have no application, and the practices of punishing and blaming, of expressing moral condemnation and approval, are really unjustified.

while the opposing optimists believe that these concepts are compatible with that thesis. A third party, whom Strawson calls genuine moral sceptics, think that 'the notions of moral guilt, of blame and moral responsibility are inherently confused', and that the difficulty of combining them with determinism only illustrates this fact.

Strawson himself finds the notion of determinism itself too confused for any of these positions to be satisfactory, and I am sure that he is right. But he sees that all of them contain elements that must be taken seriously. His central thesis is that our personal responses to each other's acts (responses typified by resentment) are an essential part of our species-specific constitution, and that the crystallization of these attitudes in morality is an integral consequence of our combining – as, again, we are species-specifically bound to do – into social groups. So, though we need to correct many confusions and excesses of moral thinking, the idea of throwing it out makes no sense. Accordingly, whatever difficulties arise from trying to combine it with determinism must be faced and dealt with by thinking harder. The current two-party system on this issue is not workable. Strawson says:

> Optimist and pessimist misconstrue the facts in very different styles. But in a profound sense there is something in common to their misunderstandings. Inside the general structure or web of human attitudes and feelings of which I have been speaking, there is endless room for modification, redirection, criticism and justification. But questions of justification are internal to the structure . . . The existence of the general framework of attitudes itself is something we are given with the fact of human society. As a whole, it neither calls for, nor permits, an external, 'rational' justification. Pessimist and optimist alike show themselves, in different ways, unable to accept this.[2]

Optimists must therefore resist the temptation to scale down morality reductively – for instance, by treating it simply as a means of social control – in a mistaken attempt to justify it by bringing it inside a convenient deterministic pattern. Provided, then, that we

> remember this and modify the optimist's position accordingly, we simultaneously correct its conceptual deficiencies and ward off the dangers it seems to entail, without recourse to the obscure and panicky metaphysics of libertarianism.[3]

WILLIAMS ON MORAL LUCK

Against this kind of position, Bernard Williams has developed a sceptical attack directed in some degree against the whole conception of moral freedom and expressed primarily in his paper 'Moral luck'.[4] Williams thinks there is something radically confused in our whole notion of moral freedom, something that makes it impossible to state it coherently. Attempts to do so fail, he says, not just because the topic is difficult but because the thing cannot be done. 'The attempt is so intimate to our notion of morality, that its failure may rather make us consider whether we should not give up that notion altogether.'[5]

I do not think it is at all clear what that giving up would mean. I have discussed this attack of Williams's elsewhere,[6] and I do not want to do so again in detail, but we do need to attend here to the moral campaign which seems central to this and similar attacks on the notion of freedom. Essentially, that campaign springs from horror about misuses of the concepts of blame and punishment. That horror is, of course, fully justified. Blame has indeed constantly been used, throughout human history, to license uncontrolled callousness and cruelty. And it is true that recent extensions of causal reasoning to cover some parts of human conduct have helped to limit this misuse. Gradually, legislators and lawcourts began to admit that insane or subnormal offenders had not been free to avoid committing their offences. It was natural, therefore, that reformers from Bentham on (and including Skinner) should propose continuing that extension to its limit and dropping the notion that anybody could ever avoid offending.

This project too I have discussed elsewhere.[7] Fairly obviously, it is one of many attempts to get rid of an abuse by ditching the thing abused. That thing, however, is unfortunately not dispensable. No blame would mean no credit and no ownership of actions – no possibility of distinguishing who did anything. Moreover, this change affects, not just the past, but the future. Williams deals entirely with guilt, remorse and regret about the past. He also concentrates largely on third-party cases – on our opinions about other people's responsibilities, opinions which, no doubt, we often need not form. He stresses that we should remember that these people may have been in a tragic bind, that perhaps they could not help what they did. But the categories of freedom exist primarily to help us think about our own way forward, about what we ourselves, and other people, should do next and thereafter. For that purpose, it is absolutely vital for us to distinguish what we can help from what we cannot. And, bad though we may be at doing this, muddled though our accounts of the situation may be, we have no choice but to press on with them and to make them better.

REMEMBERING FATALISM

To sum up – There are indeed serious questions about just *how free* we should take ourselves to be. There have been terrible excesses in both directions. Is it plausible that the excesses produced by the abuse of blame quite outweigh the fatalistic excesses produced by dishonestly including oneself in the causal process – by bad faith? To suggest this is, I think, to forget the appalling effects of fatalism. For example, it neglects the huge political consequences of pretending that we are unable to influence public affairs. If we want a more personal case, there is a good one in *Les Liaisons Dangereuses*, where the Vicomte de Valmont allows his spiteful ex-mistress to finesse him into treating his own behaviour as mechanical and predictable, thus making himself her tool. On her advice, he sends to the woman he has just seduced a letter which treats his behaviour as unavoidable. It begins:

> One grows weary of everything, my angel, it is a law of nature, it is not my fault.

> If therefore I am weary of an adventure which has wholly preoccupied me for four mortal months, it is not my fault.

> If for example I had just as much love as you had virtue (and that is surely saying a lot) it is not astonishing that one should end at the same time as the other. It is not my fault.

> From this it follows that for some time I have been deceiving you, but then your pitiless affection forced me, as it were, to do so! It is not my fault.[8]

Valmont means his letter to make him, as it were, a registered automaton, to clear him somehow of responsibility, but of course it does nothing of the kind. As his ex-mistress points out,[9] he has not really managed to become a mechanism. Instead, his action was in fact a deliberate move to save his vanity, because he was ashamed to appear carried away by his quite genuine love, and (as she adds) he has been at least as weak in letting himself be carried away by vanity as he would have been if he had let love rule him.

This story, like many in that most disturbing novel, is of course unusual in that the matter is made so explicit. Like French classical tragedy, it uses an unrealistic psychological convention where motives appear openly in isolation. The novel, however, is none the less powerful for that, because what is said through that convention is surely a matter of common experience. People do extremely often make, less openly, Valmont's kind of move. The last section quoted here plays what Eric Berne describes as the game of *See What You Made Me Do*, with its subsidiaries of *You Got Me Into This* and *Why Does This Always Happen To Me?* – all of which are everyday matters.[10] Over all, it is surely not at all obvious that this sort of fatalism is any less damaging in human life than the excesses of blame are.

FATALISTIC DRAMAS

The central issue here, then, is not about determinism. It is about fatalism, about the impression that we can make no difference to what happens. If we believe that we are powerless we become so. This sense of powerlessness does not follow from scientific determinism. It is merely a common, seductive dramatization of it, which proceeds by telling us that we are in the grip of ineluctable forces.

Mr Jones, who has always been a great reader, might well read *The Selfish Gene* on his deathbed, and it might convince him that he was not really free because he is a vehicle exploited by his selfish genes and memes. He might feel helpless, and this could even help to kill him. But if so, it would do so by a confusion. Like a rope-dancer infected by the spectators' nervousness, he would lose confidence because he had responded to a mood, not to an argument. The real future – the destiny that 'is to be' – could just as well be that he will decide to think harder, will reject the argument and the mood with it, and will continue his efforts and survive.

The thought that we are composed of particles viewed by physical scientists as forming a deterministic system is primarily their business, not ours. When it is quietly and objectively presented, it does not alarm us. The trouble starts when someone decides to ham it up with talk of ruthless, exploitative mechanisms. This talk of mechanisms, if not carefully controlled, always suggests a manipulative designer and user, a puppet-master who rules our lives because he has set the whole thing in motion. Such talk is a degenerate descendant of the most odious representations of Calvin's God.

When people say that 'the organism is only DNA's way of making more DNA'[11] or that 'we are survival mechanisms – robot vehicles blindly programmed to preserve the selfish molecules known as genes',[12] this melodramatic mode is unmistakable. Fatalism tells us, not just that the future is fixed, but that we are helpless in the hands of alien forces. Against this lurid background, our own purposes become somehow unreal. Now that we are no longer supposed to have actual malign gods, this illusion works most easily against either a biological or a social-science background. While we are talking physics and chemistry, fatalism does not easily get a foothold, because purposes are clearly irrelevant to those subjects. A strictly physicalist universe might be chilling, but only in a negative way. It may certainly seem to suggest the absence of purpose from the cosmos. But even that suggestion is really beyond its brief. Properly speaking, physics simply has nothing to say on such topics.

Biology, however, can much more easily be twisted to lend colour to melodrama. This sociobiological rhetoric makes its effect by inflating the modest concept of evolutionary function into a personified competitor with ordinary individual purpose, a competitor who now triumphantly manipulates it. This superpurpose is then given an owner – not a straightforward old-fashioned deity but, more creepy still, a kind of plotting goo at the heart of our own body-cells. We emerge as the pawns of our own tissues, pawns deluded into thinking that they are the players.

WHY METAPHORS MATTER

The damage done by this drama cannot be cured by putting occasional notes in the books saying that the language is meta-phorical. Of course it is metaphorical. The question is, what does the metaphor convey? Metaphors are not just cosmetic paint on commu-nication. They are part of its bones, crucial members in the structure of thought. Science itself is packed with examples that show this. For instance, it is clear how much influence the metaphor of 'selection' had on Darwin's thought, and what a deep effect the imaging of particles, first as billiard-ball-like parts striking each other in a 'mech-anism', and then as waves or solid items, has had on physical theory.

These images are not just loose, optional devices for explaining physics to outsiders. They have always been essential parts of the conceptual system. They work as pointers towards particular ranges of theoretical possibilities, ranges which, so far, are only seen in outline. Those pointers can be immensely useful. But in following them, the first need is always to remove irrelevant ideas which the metaphor is liable to suggest.

All metaphors have their misleading features. In order to guard against them, it is essential not to rely blindly on a single image.

Sensible thinkers use one to correct another, as Einstein constantly did, and as physicists have done in the case of waves and particles. In fact, people who find their thought being dominated exclusively by a single image ought always to become suspicious, to look for the limitations of that image, and to warn their readers about those limitations. Darwin sometimes tried to take these precautions about selection, but, as has since been well noted, he did not grasp the full force of that powerful metaphor.[13]

SOCIOBIOLOGICAL TROUBLE

This suspiciousness is obviously even more vital when the irrelevant suggestion is an already existing doctrine with notorious faults, such as the Social Darwinist reliance on psychological egoism. Writers who, instead of taking these precautions, actually choose their technical terms (such as *selfish*) from the doctrine in question and shape their language to suit it, make it clear that they are not just using careless rhetoric but are backing that doctrine as part of their message. Readers will then, quite correctly, read the metaphor as conveying that doctrine, and, if they notice the disclaimers, will disregard them as humbug.

It is worth while to notice how this fatalistic reduction works in a couple of examples from the rhetoric of sociobiology;

The brain exists because it promotes the survival and multiplication of the genes that *direct its assembly*. The human mind is *a device for* survival and reproduction, and reason is just one of its various activities.[14]

Beliefs are *really enabling mechanisms for survival*. Religions, like other human institutions, *evolve so as to* enhance the persistence and influence of their practitioners . . . Thus does ideology *bow to its hidden masters* the genes, and the higher impulses seem upon close examination to be metamorphosed into biological activity.

Similarly, in a passage already quoted:

Human behaviour – like the deepest capacities for emotional response which drive and guide it – is the *circuitous technique* by which human genetic material has been and will be kept intact. Morality has no other demonstrable *ultimate function*.[15]

It is surely instructive to see how these wild expansions are combined with a stern, reductive tone which calls on us to give up superstition in favour of reality. It seems that we have been mistaken in supposing that we were reasoning, since – really – we were only being manipulated so as to promote the survival of our genes.

Now the small biological point being made here is sound. It does seem likely that our elaborate brain would not have developed if it had not, on the whole, promoted survival. That, however, is a truism about all our organs. When we ask for the function of any particular organ, this general truism is irrelevant. We are asking about the special advantage this particular organ gives to its owners. That question has to be considered from the owners' point of view, with regard to their particular needs. They – the whole organisms – are the only beings whose aims can come into the matter.

Certainly these organisms are themselves 'products of evolution'. But this is not like saying that a Bugatti is the product of its firm. Bugattis only exist to serve the purposes of their designers. Organisms do not have a designer. Their own DNA, which is a part of them, cannot act as one. Neither could the abstraction called 'evolution'. In biology, evolution is not considered as a vast designer, but merely as a large process showing certain general tendencies. Talk of evolutionary function gets its meaning only in relation to those tendencies. If we move from this point of view – if we start claiming to have discovered the real purpose, the 'ultimate function' of any phenomenon – then we have left biology and begun to talk metaphysics. Almost certainly, indeed, we have begun to talk religion. And these are kinds of talk that proceed by very different rules.

Can sociobiologists really have been so naïve as to make this mistake? It certainly does seem remarkable that they should take their obviously misleading metaphors so literally. Yet their writings often explicitly stress that they do so. Today, two factors make this kind of mistake peculiarly easy and tempting. One is the familiar, constant use of the machine model in the sciences – a model which, of course, is not supposed to include a designer, but which, by its very nature, always tends to suggest one. The other, also very familiar, is the pattern of *psychological* reduction which we noticed before as applied to motive. It is the pattern by which we say 'all this patriotism is really only avarice – or vanity – or a means to a knighthood'.

The knowingness expressed here feels quite like the knowingness with which we remind each other that we are but dust, or that a human body is just five pounds' worth of chemicals. The first – psychological – kind of knowingness is displayed in the idea that human life is wholly determined by selfishness. The second – physicalist – one produces the idea that it is wholly determined by DNA. In both these moods, sociobiologists have the impression that they are just fulfilling a single important reductive function, namely, debunking human pride. They tend, therefore, not to distinguish between the two moves. And they conclude that the opposition they get is simply the predictable response of outraged human conceit. They never notice that their expansions outweigh their reduction.

CONCLUSION

Our discussion, throughout this second part of the book, has centred on the mistakenness of trying to find a single 'fundamental' form of explanation for very complex matters such as human activity. We have considered various kinds of reductive campaign that have been inspired during the last two centuries by the hope of reaching that elegant pattern, campaigns which are still being vigorously carried on today. As we have seen, the trouble is not that the wrong candidate was chosen as fundamental. No form of thought is fundamental in that way. No such single candidate could succeed.

The elegant, monistic pattern is simply unusable. Instead, explanation of complex things has to proceed pluralistically and convergently, not by competition between specialists but by using many different ways of thought that converge on the topic and co-ordinating their findings. This approach is not less rational than reductive monism. It is more so, since it works.

Attempts to people the world solely with objects and to leave no subjects to observe them are not very realistic. A conception of 'objectivity' which ignores the fact that we have to talk about subjects, and can to some extent do so objectively, is therefore useless. This much has often been pointed out before. What has less often been stressed, however, is that conscious subjects are, essentially and by their nature, not just observers but *active agents*. This means that ways of thought which ignore these subjects do not just miss the details of introspection or pain or colour-vision, though those can be important enough. They also miss the whole language of action, the concepts used for discussing what to do.

That is surely the reason why a gap has opened up, not just between supposedly scientific explanation and all subjects of direct human concern, but more deeply still between rational explanation and morality. It is why moral talk, which belongs to the language of action and is central to it, has been seen as falling outside rationality and therefore undiscussable. It is why the kind of freedom that morality presupposes has seemed inconceivable. It is also, then, why our condition as moral beings has seemed to cut us off from our evolutionary origins, making it seem incomprehensible how morality as we know it, morality in anything but the crudest, most reductive sense, can have originated. That, then, is the subject to which we must move in Part III.

Part III
THE SOURCES AND MEANING OF MORALS

9

AGENCY AND ETHICS

WHAT IT MEANS
NOT TO BE A MACHINE

HAVING GLANCED, then, at the ways in which reduction currently works, let us go back to our first question. Why is an evolutionary origin a threat to us? In what way is it dangerous to form part of the natural world? If – as seems plausible today – what we fear is loss of freedom, just what kind of freedom is it that we need? Can our primate history spoil it?

People picture freedom in a thousand ways, varying with the kinds of control that gall them. Some concepts of it are very complex. But the kind that seems to be in question here can, for a start, probably be outlined fairly simply. It centres on our being real individuals, agents rather than just pints of water in a river or lava in a volcano. It necessarily involves the ambitious notion that we are not wholly continuous with the scene around us. We are somehow separated off in a way that gives us some kind of power to choose, to direct our own course.

The contrast that commonly comes to mind if we want to make this idea plain is that we are not cogs in a machine. This fits our present enquiry well, because, throughout the history of modern science, machines have been the standard model for describing the natural order. The seventeenth century's fascination with fine clock-work, repeatedly strengthened by the growth of later machinery, has given such patterns of thought huge power. Machine metaphors are so familiar that theorists often forget that they are metaphors at all. They treat them as mere literal descriptions, as we saw in Barrow and Tipler's remarks on human computers in Chapter 1, pp. 9–10.

It is not surprising, then, that this image has been used over-confidently, indeed often wildly. For more than a century now, scientists themselves have been pointing out the misleading consequences of this. Physicists in particular have issued strong warnings about the limitations of mechanistic thinking. As David Bohm put it,

> While we do not wish to suggest that the analogy between electrons and living beings is complete, we do wish to emphasize that it goes far enough to show that physics has really abandoned its earlier mechanical bias. Its subject-matter already, in certain ways, is far more similar to that of biology than it is to that of Newtonian mechanics. It does seem odd, therefore, that just as physics is moving away from mechanism, biology and psychology are moving closer to it.[1]

Mechanism is, however, one more very persistent imaginative habit. Many people still see it as essential for a scientific approach, and assume, too, that it is extendable without change to social and psychological problems.

Since the kinds of questions that arise here are very different, this is an odd assumption. As Rom Harré and his colleagues put it;

> When a volcano erupts, vulcanologists explain the event by reference to a chain of causes ... The mountain is a mere passive component in the 'mechanism' of an eruption ... When the conditions are right, each step in the chain of causes and effects leads inexorably to the next.

The volcano, in fact, is just a hole through which stuff pours. It does not do anything. It differs from individuals who do act in many obvious and important ways. Indeed, as these authors say;

> Causal mechanisms in the physical world seem to be characteristically different from the mental processes which underlie thought and action, and which are involved in how we interpret our feelings. *At first sight one would think that a great deal of effort would be put into exploring the similarities and differences between the causality typical of physical processes and the ways people manage their actions, develop their thoughts and display their emotions.* But psychology has not developed like that.[2] (Emphasis mine)

What, then, does the machine metaphor actually mean? Its central point seems to be continuity with one's surroundings. Though cogs are separate objects which can be distinguished from one another, they are not distinct subjects and therefore not properly individuals. They are inactive material in the hands of their designers and users. But it is, of course, possible to get doubtful about the force of this distinction. Might the human designers and users themselves also be cogs? Might we all, in some sense and at some hidden level, be ourselves passive? Is everything really just one vast machine?

The difficulty about this suggestion is that it does not make much sense without further designers and users to give the proper contrast. Without them, there is no active element in charge, and so no real passivity. When this model began to take shape in the seventeenth century, it made easy sense because it presupposed God as the active designer. But when it began to be used to get rid of that active, external figure its point became increasingly obscure.

When the metaphor of mechanism is applied to physical processes such as those in volcanoes, its chief point is that they are regular and can be understood as continuous with their environment in the sorts of ways by which designers and users understand their machines. The point of extending it to cover intentional human action has to be to suggest that this action too should be understood in the same way, rather than by using the many more convenient, more specific ways of understanding intention which humans from their earliest days have taken the trouble to develop.

This raises a sharp puzzle about the position of the understanders themselves – the scientifically qualified people who use these mechanical principles. These people are clearly not just seen as one more set of cogs. Their place seems to be more that of a mechanic dealing with a machine. So it begins to look as if they are really agents somehow standing right outside the processes that they are studying. They appear, in fact, to have taken over the place of God in the older model – not indeed as creators, but as a kind of Providence. They function as the only real agents around, those responsible for running the human machine, active beings who are free to direct its course because they understand its workings.

THE ATTEMPT TO
ACCEPT PASSIVITY

This peculiar status for social scientists was never clearly stated. But the whole literature of behaviouristic psychology implies it. B.F. Skinner devoted his long and influential career to attacking existing ideas of individual freedom and replacing them by a mechanistic model in which psychologists would run the workshop that would service the system. He constantly attacked the idea of activity or 'autonomy' as a mere superstition:

> In the traditional view a person is free. He is autonomous in the sense that his behaviour is uncaused. He can therefore be held responsible for what he does and justly punished if he offends. That view, together with its associated practices, must be re-examined when a scientific analysis reveals unsuspected controlling relations between behaviour and environment . . . Science has probably never demanded a more sweeping change in a traditional way of thinking about a subject, nor has there ever been a more important subject . . . The direction of the controlling relation is reversed: *a person does not act upon the world, the world acts upon him.*[3] (Emphasis mine)

On these new, supposedly scientific principles, then, the inner thoughts and questionings that seem to shape our choice are illusory. They do not actually affect our behaviour at all. They are, as Skinner repeatedly insisted, just a side-effect, a trivial consequence of our behaviour, an insignificant froth on its surface:

> Any feelings which may arise are at best by-products. Our age is not suffering from anxiety, but from the accidents, crimes, wars, and other dangerous and painful things to which people are so often exposed . . . *We can follow the path taken by physics and biology by turning directly to the relation between behaviour and the environment and neglecting supposed mediating states of mind.* Physics did not advance by looking more closely at the jubilance of a falling body, or biology by looking at the nature of vital spirits, and we do not need to try to discover what personalities, states of mind, feelings, traits of character, plans, purposes, intentions, or the other perquisites of autonomous man really are in order to get on with a scientific analysis of behaviour.[4] (Emphasis mine)

When Skinner wrote this in 1970, the difficulties of his programme had already become clear. They have since proved so crushing that shrewd theorists no longer promote it. No way has been found to explain plausibly how people could act at all on this pattern, even in doing rather quiet actions such as writing books. How, for instance, would environmental pressures so affect Skinner, or his opponents, as to produce passages of prose like this without any mediating processes in their minds? What kind of causal laws could predict the contents of future books from outside circumstances, without reference to the thoughts of the authors and readers? And if scholars themselves do not live like this, why should they suppose that other people do?

There is no way in which serious thought could be supposed to be so passive, or in which ineffective thought could be worth attending to. Unmistakably, then, theories like this are propagandist theories about other people. They are meant to be about the rest of us, the psychologists' subject-matter, the common herd of mankind. They are not, and cannot be, about the theorists themselves, who must still be free agents if they are to play their part in controlling the world.

Besides these difficulties about content, however, the tone was also significant. Skinner's triumphalist style, his gross over-simplification, his open contempt for all thought and all professions except his own, repelled many people, both learned and unlearned. But many others have found them attractive, seeing them as a sign of justified confidence. Both the style and the content of this behaviourist message have accordingly had huge influence and still do. They do not owe their force to Skinner's being an original thinker, but to the fact that he was not. Like Wilson, he expressed dreams that were already in the air, dreams which have by no means gone away. He sketched out openly and in strong colours a myth that was already strong as a half-understood aspiration.

There is a strange paradox here. This myth, which seems to remove the idea of human power altogether, owes much of its charm to its offer of power to the experts. It promises that those who will study a certain science – psychology – can thereby acquire an almost magic ability to manipulate human behaviour. They will no longer need to allow the behavers any choice. They will themselves become free agents endowed with hitherto unknown control over others. But the myth's emotional tone can also gratify an ambivalent response to power on the part of those manipulated, a depressed masochistic desire *not* to be free, a fascination with passivity, which has often found expression during the last century in blurring the gap between people and machines. Skinner made this myth look intellectually respectable in an ideology that also has the appeal of seeming to simplify life amazingly.

TRYING NEURO-SCIENCE INSTEAD

Other people, however, are indeed appalled by this same colourful myth, and see that it has been a prime source of 'anti-science' feeling today. Accordingly, more sophisticated reducers have recently been turning away from explanations of human conduct centring on environmental causes towards ones that combine neurobiology with information theory. These approaches, however, still often agree with Skinner in dismissing the concept of individual agency as a mere dispensable piece of folk-psychology, a colourful everyday notion which must eventually give way to the more scientific idea of an impersonal process. On this view, scientifically speaking, at a deep level, people do not really act. There are just events. People, like volcanoes, are holes through which events flow.

Unlike Behaviourism, however, this physicalist approach finds the most relevant events to be those within the meat of these people's own heads and nervous systems rather than in outward behaviour. And it typically considers that these events are quite literally identical with the mental events that accompany them. The two are not separate, so no difficulty can arise about how to relate them.[5]

In some ways, this approach has certainly been a great improvement. It has been intended as a reaction against both the brutality and the unreality of Behaviourist theory. By locating the most relevant events inside rather than outside each person it shows more respect for individuality. And by being prepared to treat conscious experience as somehow a legitimate aspect of these events, rather than as a mere meaningless by-product of them, it tries to show more regard for the subjective viewpoint.

That is good. But, if one has moved so far, it is hard to see why one should not move further, and indeed how it is possible to stop at this point. The trouble is that conscious experience, though it may very well be closely linked to events in the nerves and brain, reaches us in a quite different way from them and does not share their structure. What is seen from the subjective viewpoint has to be described and understood in entirely different terms.

Even if there is at some deep level a pattern connecting the two, a pattern which some distant observer – God again – would be able to detect, no such pattern is revealed to us. We have no reason other than optimism for believing in it, and certainly no reason for supposing that it gives physical explanations any priority over mental ones. From where we stand, as opposed to where such an omniscient witness might stand, the discontinuity between the subjective and the objective viewpoints is a real one. Accordingly, to say, as Paul Churchland does, that 'conscious intelligence is the activity of suitably organized matter'[6] is not to begin an explanation. It is not at all like saying that the morning star is the same as the evening star, a remark whose sense can be explained quite easily. Instead, it simply proclaims a huge act of faith.

Of course it is true that researchers have lately made huge and fascinating discoveries about the neural processes involved in thought and perception. They have evidently hoped that, by filling in this story – by tracing a continuous neural process – they can build up a context which will make the transition to conscious perception itself equally smooth and unremarkable, thereby solving 'the problem of consciousness'. But this is simply building a smooth road to the edge of the Grand Canyon. What comes next is a change of a totally different kind. The Grand Canyon is still part of the natural world, but roads cannot be bulldozed across it. There is no possible kind of build-up that could make it less of a jolt to move from talking about how nerves work to the first-person experience of – say – a sudden toothache or a blaze of light. This move is not a further stage of the same process. As Raymond Tallis puts it:

> It has been mistakenly thought that these kinds of observation – which merely describe how one form of energy is transduced into another – contribute to an understanding of how we perceive the world. The hidden assumption – absurd as soon as it is made explicit – is that the process by which energy . . . gives rise to experience is somehow analogous to that by which one form of energy is transduced into another.[7]

But, as he points out, this last process is actually the quite different one which raised questions in the first place; it is the *explanandum*. Transduction has nothing to say about it. 'There are many transducers that are not sense-organs – for example photo-electric cells – so transduction itself is not sufficient to create sensations.' How, then, (asks Tallis) can it have been possible for theorists to slur over this glaring gap? He replies that they do it by constantly confusing the distinct languages that belong to the two distinct viewpoints;

> The power of neuromythology resides in the subtlety with which it juggles descriptive terms. Neurophysiological observations seem to provide an explanation of perception only because those observations are described in increasingly mentalistic terms as one proceeds from the periphery to the centre of the nervous system . . . *As a nerve impulse travels along an afferent fibre, it also propagates from one page of*

101

Roget's Thesaurus to another ... It leaves the world of 'energy transformations' and enters the world of 'signals' until, two or three feet and two or three synapses later, it has become 'information' ... No explanation whatsoever is offered for how this happens – and yet it cries out for explanation.[8] (Emphasis mine)

THE GAP THAT DOES NOT GO AWAY

This is, in fact, just one case among many where inattention to the workings of language produces, not just obscurity, but a total misdirection of enquiry. At a learned level, this kind of mystification is burgeoning at present because theorists now feel quite strongly that they need to do some kind of justice to consciousness. Yet they are still bound by outdated reductive conventions that stop them giving it anything like the space it needs.

Taking consciousness seriously involves accepting the crucial, if regrettable, fact that there is a real discontinuity between the inner and the outer standpoints. It forces us to see that the connections that can be established here must, so to speak, be reached sideways and convergently, by studying the wider context within which these different standpoints are related. They cannot be found by simply pushing on in the direction in which one happens to have started. And to look at consciousness in this wider context must include accepting the everyday notion of individuals as real agents. It is useless to invent theories which marginalize the fact that ordinary life – including, of course, the business of enquiry itself – still depends radically on conceptions of this genuine individual activity. That situation must somehow be understood, not suppressed.

This everyday notion is not as strange as it has often been made to sound. As we shall see later, it does not involve the wildly excessive claims to independence that have been made by moralists such as Sartre.[9] It does not call for Descartes' bold metaphysical view of the soul as a substance that can exist on its own. Nor does it call for some peculiar component within us, such as the will or the intellect, to be separated off and act freely while the rest remains enslaved. What it does call for is that each one of us can, to however slight an extent, understand our own position and our choices and thereby act individually, as a whole.

We know very well that this kind of individuality is sharply limited. We know that we have been formed from earthly materials by earthly processes and that we still constantly depend on them for our whole being. We know that we are separate, harmonious beings only in the limited sense in which other organisms are so. Yet this situation as an organism does not put us in the position of a volcano. Organisms are unities in a much stronger sense than volcanoes are, and humans are equipped with faculties that enable them to unify themselves, if they please, considerably further than other organisms.

This limited sense of wholeness and separateness is an indispensable condition of our living the only kind of life that is possible for our species. It makes no sense to talk of dismissing so deep a structural factor as an illusion. Instead, the only question is, how to interpret it. Human cultures have indeed often put unduly bold interpretations on that natural sense. And our own culture, in particular, has grossly exaggerated the degree of independence that individuals have, their separateness from other organisms, and also their degree of inner harmony.

But these exaggerations do not affect the more modest facts that underlie them. Whenever people have to take decisions, the language of agency has to be used, and the reasons why it had to be invented constantly become obvious. The language of impersonal process, by contrast, can scarcely be used at all for many important aspects of human behaviour and, when it is used there, it often serves only for fatalistic evasion.

There is no way of living that will not constantly require us to ask who did something or who will do it. Distinctions between cogs and the people who handle them have huge practical importance. So have distinctions between people who do something on purpose and those who do not. Unless, then, we believe that ordinary life is somehow less real than theory – a metaphysical view which is not now popular – these concepts cannot be junked or demoted. They are not just provisional folk-psychology. They are the right tools for the job.

VALUES ARE CENTRAL FOR ACTION

What notion of activity or agency do we then need? There is, of course, a whole category involved here, not just a single notion. We need a whole distinctive tool-box, a set of maps, a batch of concepts that are adapted to dealing with people rather than just with things. To name just a few of its most obvious divisions – there is the concept of acting deliberately rather than casually or accidentally, with many subdivisions for different kinds of casualness and accident.

Within deliberate action, there is the idea of having reasons for acting, of consistency or conflict between those reasons, and of ways to reconcile them. There is an invaluable set of concepts for discussing effort – strenuous or otherwise – and lack of effort, ease and difficulty, co-operation and obstruction, success and failure. There is, too, the essential notion of choosing better or worse. If more than one person happens to be involved, there are also notions of responsibility, of approval or disapproval, of being praised or blamed, rewarded or punished, excused or not excused for one's choices. This tool-box is, in fact, necessarily an evaluative one. It is a kit whose use naturally leads people to develop a morality.

Designers and users of machines, whom we mentioned earlier, always need some items from this tool-box. They need these concepts even at the very simple level of trying to find the best way of carrying out an obvious and unquestioned aim. People who choose between various ways of achieving such an aim are already beginning to evaluate, which is something that cogs do not do. But of course the more interesting and serious kinds of evaluation begin when conflicts of aims arise.

There are bound to be such conflicts at even the simplest level of human life, and indeed before it. People ask, for instance, not just 'how shall we build this enormous fish-trap?', but also, 'do we need it at all? Might it perhaps be better to make a smaller one and save some of the logs for the houses of the people who carried them in? Or again, should we perhaps pile some of them up inside the cave so that we can get at the ceiling and paint antelopes on it, as we were wanting to do? What, in fact, do we really want when we think about it? What will it be best for us to do?'

At this point practical thinking is needed, thinking which unmistakably does determine action. Instantly, theories like Skinner's collapse, as may be seen in the hilarious chapter on 'Values' in *Beyond Freedom and Dignity*. After asking, at the start of that chapter, several highly relevant questions such as 'for whom is a powerful technology of behaviour to be used? Who is to use it? And to what end? . . . What, in a word is the meaning of life?'[10] Skinner delivers his answer:

> If a scientific analysis can tell us how to change behaviour, can it tell us what changes to make? *This is a question about the behaviour of those who do in fact propose and make changes* . . . The reinforcing effects of things are the province of behavioural science, which, to the extent that it is concerned with operant

reinforcement, is a science of values . . . Relevant social contin-gencies are implied by *'You ought not to steal', which could be translated,'If you tend to avoid punishment, avoid stealing'.*[11] (Emphasis mine)

But people who are wondering what they really want, or what it might be best to do next, are not looking for statistical information about their own past behaviour. They may need these and other facts, but only as raw material for their decision, not as substitutes for the decision itself. They may, too, be looking for the best means to a familiar end. But they may just as easily be facing a conflict of ends. Such clashes raise a quite new kind of question – the kind which is the start of moral thinking, and also of the special kind of freedom that is our present business.

PRIORITY PROBLEMS

How could it conceivably be unscientific to take notice of these clashes, and of the kind of organized thinking that handles them? What really is unscientific is the refusal to recognize honestly the complexity of the topic we are studying. If the idea of 'cause' is still to be used here, and is not to be a mere sham, its sense will at least need to be carefully widened, an option which Rom Harré discusses:[12]

Perhaps we should treat decisions, plans and so on as a special category of causes. The difference has to do with the actor's relation to the 'program' of his or her action. A real actor could have done otherwise. But when we are thinking in causal terms it seems difficult, if not impossible, to justify that important qualification. Such a qualification has the further consequence of leading us to think of what the actor did, or didn't do, in moral terms. *The old psychology tried to study human action within a causal order, while the new psychology tries to reach a scientific understanding of human life within a moral order or orders.*[13]

It does so because all our conceptual schemes – including the scientific ones – get their sense within a moral order, a comprehen-sive practical mapping of values and standards by which we direct our lives. All theoretical thinking proceeds within this matrix, not as an up-to-date substitute for it. The ideals that guide scientific practice –

ideals such as impartiality, truthfulness, thoroughness, parsimony and the rest – are parts of that moral order. But they are not the whole of it, and they are certainly not an independent private company competing with it.

If freedom and morality are indeed closely linked in this way, it is perhaps a rather paradoxical fact that the first effect of freedom should be to put us under these new constraints. Our freedom is exactly what gives us these headaches, what makes possible this moral thinking, this troublesome kind of search for priority among conflicting aims. By becoming aware of conflict – by ceasing to roll passively from one impulse to another, like floods of lava through a volcano – we certainly do acquire a load of trouble. But we also become capable of larger enterprises, of standing back and deciding to make lesser projects give way to more important ones. That, it seems, may be why moralities are needed.

THE NEED FOR ARBITRATION

Is this, then, the origin of ethics? Is this how the whole troublesome moral enterprise became necessary and possible? If we ask how it began, we are of course asking two questions, not one. There is the factual question – what actually happened? This question is interesting enough, but, asked on its own, it would in principle be much like other historical questions about the origins of anything else, such as meteorites. It can, however, scarcely be asked on its own. It necessarily raises also the much more puzzling question about authority. Why and in what way does ethics now bind people?

It is the second of these two linked questions which has always caused most anxiety. In the last few centuries, theorists have tried strenuously to calm this anxiety by separating the two questions, handing them to different specialists to answer. This has a point in so far as specialized knowledge can indeed throw useful light on parts of the problem. But the anxiety that surrounds the second question persists, and it demands the wider perspective.

That anxiety is obvious in many traditional myths about the origin of the universe. These myths commonly try to explain, not just how human life began, but also why it is now so hard, so painful, so confusing, so conflict-ridden. They often tell of primal clashes and disasters, because their chief aim is to locate the sources of trouble, to discover why human beings have to live by rules which so often frustrate their desires. Sometimes indeed this kind of explanation is the primary theme of the stories. Interest in this whole topic does not flow just from curiosity, nor just from the hope that we may prove the rules unnecessary, though these are both strong motives. It arises, and it is constantly being renewed, because of conflicts within ethics, or morality, itself. (I do not think that we need any special distinction between these two words for our very general purposes here.)

Even in the simplest culture, the most fully accepted duties can sometimes clash, not only with people's wishes, but also with one another. The conflicts suffered by Hamlet and Orestes and King Arjuna are not exceptional. There is no primal state of innocence that antedates these conflicts. When they arise, people begin to need deeper, more general principles by which they can arbitrate between their rules. They must look for the point of the different commands involved, and try to weigh those points against each other. This search naturally leads them to look for some kind of inclusive explanation, for something that will make clear the point of morality as a whole, and will serve as a criterion for all the conflicts that arise in it.

WHY ORIGINS MATTER

That is why our original question is so complex. Moral rules, unlike meteorites, have an authority as well as a physical source. (Rules are not actually the whole of morality, but they are the point where the clashes tend to become most obvious.) In asking why rules matter, we have to imagine what life would be like without them, and this does raise questions about their actual origins. When we try to understand any aspect of human life which can cause great pain – death, disease, tribal divisions, war, oppressive government, painful marriage customs – we naturally look backwards, asking whether life was once free from these things. And morality is certainly among the things that can cause people great pain.

Was there, then, ever an 'unfallen' conflict-free state, a state that needed no moral rules? Was this perhaps a state where nobody ever wanted to do anything bad? Or was it one where, in some way, nothing actually *was* bad? Did people once live 'beyond good and evil' in a sense much stronger than the one Nietzsche found for these words – a simple, drastic sense that would leave the words without any meaning? If they did live in either of these states, how did they come to lose that pre-ethical condition? Can we now get back to it? Are we still capable of this splendid evasion?

When people ask these questions, the connection between the factual issue and the one about authority becomes plain. Though fact and value often need to be distinguished they are not (here or anywhere) totally separate aspects of thinking. Most large and important questions have both these aspects, often closely linked. And in this case the link is obvious. Facts about the origin of something often do cast light on its nature. They can indicate what it is like now in a way that might not have been discoverable otherwise, and so help to show why it is important.

It is a reasonable thought that we would surely understand better what way of life suits human beings if we understood their nature better, and that knowing more about its source might really help here. Ignorance and confusion about how human psychology actually works, about what will best satisfy it and what can be expected of it, is indeed a strong factor in our moral difficulties. And it seems plausible that we might get light on that question if we really understood how the human race got involved with morality in the first place. A recent book puts the point understandably if a trifle wildly;

> We humans are like a new-born baby left on a doorstep, with no note explaining who it is, where it came from, what hereditary cargo of attributes and disabilities it might be carrying, or who its antecedents might be. We long to see the orphan's file.[14]

It might be more plausible to say that we have plenty of notes, but none of them seems quite reliable – a mass of tradition, theory and speculation, but little certainty. And we live in a varied and changing world in which new understandings of our nature seem strongly called for. The huge excitement that attends even the slightest new archaeological discovery about our early ancestors surely shows the strength of this hope. And for related reasons similar, though more surprising, hopes often greet reports of discoveries about the Big Bang.

10

MODERN MYTHS
— •◦• —

EDEN AND THE SOCIAL CONTRACT

UNTIL QUITE recently, this whole range of questions about the
source of morals was answered in our culture by a series of powerful
myths. Myths are not lies, nor need they be taken as literally true.
They are symbolic stories which play a crucial role in our imaginative and intellectual life by expressing the patterns that underlie
our thought. Our own culture has largely relied on two such
answers, both of them embodied in myths about actual origins. One
answer – coming mainly from the Greeks and from Hobbes –
explains ethics simply as a device of egoistic prudence. Its origin-
myth is the Social Contract. It sees the pre-ethical human state as
one of solitude. As Rousseau put it in his early *Discourse on the
Origin of Inequality*:

> Having no fixed habitation and no need of one another's
> assistance, the same persons hardly met twice in their lives, and
> perhaps then without knowing one another or speaking
> together . . . They maintained no kind of intercourse with one
> another, and were consequently strangers to vanity, deference,
> esteem and contempt; they had not the least idea of *meum* and
> *tuum* . . . The imagination, which causes such ravages among
> us, never speaks to the heart of savages, who quietly await the
> impulses of nature.[1]

By this account, the primal disaster was that people ever began to
meet at all. Once they did meet they were bound to clash, so that
unless something was done the state of nature must be what
Hobbes had called, 'a war of every man against every man'.[2]
Morality was invented in order to impose an armed peace. Rousseau
insisted that there was no actual war; people had not been positively

hostile to each other before the Contract.[3] But he agreed with Hobbes that survival itself, let alone social order, had only become possible because the sheer dangers of anarchy had forced beings who were natural solitaries to make a reluctant bargain. This story was, of course, usually seen as symbolical, not as literal history, but that did not lessen its power.

The other account, the Christian one, explains morality as our necessary attempt to bring our deeply imperfect nature into line with God's will. Its origin-myth is the Fall of Man, a choice which has rendered our nature radically imperfect in the way described – again symbolically – in the Book of Genesis. It is not surprising that these two simple accounts have been popular. Simplicity itself is always welcome in a confusing world, and each of them does contain some real insights. But simple accounts cannot explain complex facts, and it is clear that neither of these sweeping formulae can really deal with our questions. The Christian account shifts the problem rather than solving it, since we still need to know why we should acknowledge God's authority. Christian teaching has of course plenty to say about this, but what it says is complex, and cannot keep its attractive simplicity once this question is raised.

I cannot discuss further here the very important relations between ethics and religion. Though noisy disputes continue to centre on them, they are really not relevant for our present purpose. The Christian view does *not* just derive our duty to obey God naïvely from his being an all-powerful creator. If it did, that derivation would, of course, never account for the authority of morals. If a bad, all-powerful being had created us for bad purposes, his position as creator would not give us a duty to obey him. Obedience might be prudent, but prudence is not morality. The idea of God is far more than just the idea of an omnipotent creator. It crystallizes a whole mass of very complex ideals and standards that lie behind moral rules and give them their meaning. But the authority of these ideals and standards is just what we are now asking about. That non-factual question is still with us.

THE PERSISTENCE OF THE CONTRACT MYTH

During the last two centuries, 'anti-naturalist' moral philosophers have tried to separate this conceptual question about authority entirely from the factual ones which earlier thinkers considered relevant to it – to shut it off from all enquiries about 'human nature'.

Quite rightly, they have reacted violently against over-simple attempts to treat morality as a mere outcome of certain chosen natural motives, such as self-interest or pleasure-seeking or desire for power. And this resistance has pushed them, even the least religious of them, constantly in the direction of dividing minds radically from bodies.

Since Kant's time, these thinkers have stressed the autonomy of morals. They have often insisted that it is independent of all the facts. Recently they have denounced as 'naturalism' and 'genetic determinism' any attempt to find a source for it in motives innately present in our species, such as natural affection. They have expanded the notion of agency that we mentioned earlier on to a scale far larger than had ever been claimed for it before. They have used a strong language of unconditional freedom, pure spontaneous activity, a language carefully designed to exclude any reliance on innate tendencies. They have depicted human choice as something self-creating, isolated, without a source, without a past, concerned only with the future – as pure creativity.

Jean-Paul Sartre expressed these ideas forcibly, but he did not invent them. They were in the air already in his time, and though Existentialism is now forgotten as a philosophy, his strange claims evidently resonate no less strongly today. It has depressed me to notice, over the years, how students who are presented with them have still continued to find them quite unsurprising:

> Man first of all exists, encounters himself, surges up in the world – and defines himself afterwards . . . To begin with he is nothing. He will not be anything until later, and then he will be what he makes of himself. Thus there is no human nature, because there is no God to have a conception of it. Man simply is. He is what he wills . . . Man will only attain existence when he is what he purposes to be . . . One will never be able to explain one's action by reference to a given and specific human nature – in other words, there is no determinism; man is free, man *is* freedom . . . This theory alone is compatible with the dignity of man; it is the only one that does not make man into an object . . . We have neither behind us nor before us, in a luminous realm of values, any means of justification or excuse . . . Man is condemned to be free . . . One can choose anything, but only if it is upon the plane of free commitment.[4]

Notoriously, this approach does have its merits. It can be a good defence against fatalism. But its claims go far beyond that. It reckons to be a final diagnosis of the human situation. And here it is worth noticing that Sartre's proposal does not take us out of the realm of myth. It just gives a new twist to our current reigning myth, which is that of the isolated, sovereign chooser. It still treats morals as linked with facts about our nature in a quite traditional, myth-centred way, but makes a peculiar choice of facts to consider. The story with which this quotation from him begins – the story of human beings as starting life grown-up, without infancy or child-hood – is a fiction of a kind typical of myth, a fiction which can only be justified if it conveys a deeper truth. Its message is that the real human being simply is not present until it becomes wholly free and independent of others.

This diagnosis only works if we are in fact beings of a kind for whom total freedom of choice has that central importance. It depends, as much as other forms of the Social Contract picture, on our being the creatures that it describes. It only changes the emphasis within that picture. The Social Contract model itself is still dominant today, as is shown, not just by the respect given to explicit doctrines expressing it, such as that of John Rawls, but by the constant succession of less formal disputes where an appeal to freedom is seen as more or less unanswerable. Free-dom, which was always an element in the contract picture, has displaced rationality as its centre, and is now sometimes seen as its sole meaning.

For instance, in recent disputes over new techniques for promoting and directing fertility, the prospective parents' total freedom of choice is often described as a right – a consideration so central as to be irresistible. This emphasis is not found only in right-wing, explic-itly libertarian writings. 'Can we deny a mother's right to shop in the genetic supermarket for healthier babies?' asks an anxious science writer in the *Guardian*.[5] There is something strange about this because in more familiar matters, even in ones of great importance to us, we all regularly accept a great deal of restriction. We are perfectly well aware that conflicts of interest make it impossible for anyone always to get what they want, and also that getting what one wants can often work out badly. Even on some questions related to fertility itself, such as adoption, it is recognized that other points of view have to be considered as well as those of the prospective adopters. In all these familiar contexts, hard experience has shown this need to balance freedom against other considerations. But whenever any new possibility is raised, the sense that it must be used to produce unconditional freedom tends to surface again.

We come back, then, to our earlier question: why is freedom so important? Is it somehow a self-evident fact about our nature that we are above all freedom-seeking creatures? Most cultures have not thought so. Earlier versions of the Contract model itself appealed to a rationality that was conceived rather as intelligent self-interest, and this appeal was naturally very important in the campaign for democratic institutions. Reformers and revolutionaries offered to rescue people from the power of tyrannous rulers in order to make them happy and prosperous, so that they could then fulfil their various other purposes. They did not offer them freedom as an end in itself, nor indeed total control over their own destinies. But that older, more pragmatic argument has now dropped into the background. Many champions of moral autonomy disown it. Economists still support it, along with some unreconstructed Utilitarians, but they are seen as a distinct philosophical party.

Thus the central myth of individualism has quietly changed its meaning, as myths constantly do, with shifting emphases in society, and has split into two distinct streams. Existential Man seems a very different person from Economic Man. Yet they are still much more closely linked than either of them seems to notice. (This indeed is evident from the fact that neither sub-species easily accommodates a corresponding Woman.) The myth itself – the myth of the original isolated, independent chooser needed for the Contract story – persists. It still provides the main image that we in the West are supposed to have of our moral nature. This becomes particularly clear at times when evidence surfaces for facts which do not easily fit it – in particular, for facts about our deeply social nature. Such occasions cause excitement, anxiety, and a hasty rush of theorists to the pumps to disprove the facts or to interpret them in some safer way.

INCREASING DISCOMFORTS
OF THE MINIMAL SOUL

About the mind–body relation, Sartre's picture is of course radically separatist. Though he was a campaigning atheist, Sartre seems to have followed Descartes readily in splitting the essential self off entirely from the physical world. This division had, however, become much harder since Descartes' day. For Descartes, the mind–body gap was much larger because both the extremes were smaller. He saw the soul as a substance consisting of pure consciousness, and matter as wholly inert stuff, alien to life. He also conceived science as essentially just an extension of physics, dealing essentially with the laws governing this inert matter.

Since his time, biology and the social sciences have expanded to fill much of the gap between his rather minimal soul and his highly abstract body. And these newer sciences claim to tell us a great deal more about ourselves than was systematically known before. Accordingly, people who now see scientific determinism as threatening but who still accept it, are forced to contract the moral self much more radically than Descartes did in order to preserve it from this threat. Like householders in a flood, they keep moving upstairs, gradually losing the use of their lower floors. Kant began this process and his followers are still continuing it.

This contraction has grave consequences for the notion of freedom. More and more, what is free seems no longer to be the whole self but a distinct entity within it. Correspondingly, the factors that menace that free self – the forces from which it must remain free – seem now not to be so much those outside tyrannies which preoccupied earlier liberators, but the remaining parts of its own nature. In Sartre's picture, the will appears as embattled, an insecure ruler dominating with difficulty the alien crowd of motives that infest his (repeat his) realm. This kind of minimalist separatism ends by generating a contempt for the natural feelings quite as strong as anything licensed by Christianity.

Sartre's problem still faces many educated people today. He wanted somehow to combine belief in a number of highly sophisticated systems for understanding human conduct with his moral conviction that the human will must be seen as omnipotent and independent of all systems. That moral conviction was something he was not prepared to question. Like many metaphysicians, he built his metaphysics primarily to accommodate a moral position to which he was deeply committed, not as a means to intelligibility. So, instead of allowing that other ideals and values might matter as well as freedom, Sartre solved his problem reductively by contracting the will to an abstract, extensionless point, an empty power of decision compatible with any choice whatever. He then flatly declared that this power stood outside determinism, and defended this view metaphysically in *Being and Nothingness*. The point on which he differs from many other equally dogmatic modern separatists, such as those devoted to artificial intelligence, is simply that he took the trouble to do this. The more usual tactic today is to ignore such questions altogether.

HUXLEY AND
THE MENACE OF THE BEAST

Contemporary alarm about the relation between mind and body is, however, obviously not confined to doctrinaire separatists such as Sartre. This alarm becomes sharpest when these facts tend to link human motivation with that of any non-human species, and is then expressed, just as it was in Darwin's time, in a blank assertion that human dignity simply rules out any such comparison a priori. The details need not even be looked at. That our moral capacities are 'what separates us from the animals' is widely seen, not just as a fact, but also as a necessary claim about their value. Any doubt cast on their uniqueness is easily felt as an aspersion on the reality and importance of morality.

It is very interesting to note how deeply T.H. Huxley was committed to this pattern, and how much it interfered with his eager propaganda for the Darwinian continuity of evolution. Again and again, when he has triumphantly laid out his proofs of the physical origin of man from other primates, he feels a psychological gulf opening under his feet. How can he possibly expect people to accept such a disgusting conclusion? Thus, a couple of chapters into *Man's Place in Nature*, after insisting on the anatomical links, he pleads at some length with his readers to steel themselves for the effort of acceptance:

> It would be unworthy cowardice were I to ignore the repugnance with which the majority of my readers are likely to meet [these conclusions.] . . . No-one is more strongly convinced than I am of the vastness of the gulf between civilized man and the brutes, or is more certain that, whether *from* them or not, he is assuredly not *of* them.[6]

Huxley thinks that the only possible way of swallowing this pill is by heroically insisting that remote, scandalous ancestors cannot really compromise their descendants. Indeed, the descendants should try to make the scandal into an asset by congratulating themselves on the distance they have travelled, on the violence of the contrast:

> Thoughtful men, once escaped from the blinding influences of traditional prejudice, will find in the lowly stock whence man has sprung the best evidence of the splendour of his

capacities, and will discern in his long progress through the Past a reasonable ground of faith in his attainment of a nobler Future.

There follows a tremendous paean on human excellence, a paean intended to show that, just as mountains are not disgraced by being made of the same materials as the plains, so 'after passion and prejudice have died away, the same result will attend the teachings of the naturalist respecting that great Alps and Andes of the living world – Man.' Huxley thus invoked human pride itself to heal the wound which this disturbing news might give it. He advises it to react by simply expanding its claims to the point where it can feel incomparable and therefore proof against insult. He calls for a kind of humanism which goes far beyond species-loyalty, for a swelling species-exaltation, almost species-worship. This kind of pride-based humanism was being forged by other sages of his time, and it is still strong today. Huxley contributed powerfully to form it. It remains an effective barrier against a clearer understanding of our real position.

Huxley wielded far more influence than Darwin on the development of controversial habits, both because he loved disputes while Darwin hated them, and because he lived on much longer to conduct them. On this topic, his position differed radically from Darwin's. Huxley had begun life as a town boy with his way to make, anxious in the first place to be an engineer and studying biology only when that proved impossible. In keeping with this slant, his central biological interest was always in anatomy, and, though he was a humane man, he had no direct interest at all in animal behaviour. It plainly never occurred to him to develop any more realistic view of animals, nor to use it as a way of mitigating people's resistance to the truth about the origins of Man. Here as elsewhere, his favoured moral stance was virile, heroic and stoical. Awkward facts like these typically made him call for heroism in enduring them rather than for an effort to understand them better.

Darwin, by contrast, was a born naturalist, drawn into biology by his intense wonder and delight in directly observing plants and animals, and confirmed in his love of them by a country life. He really did regard humans as one animal species among many. Speculation about the connections between their respective ways of living was one of his central interests. As we shall see, this openness made possible much more realistic suggestions about ways to understand these connections than can be found along Huxley's path.

HOW MYTHS WORK

Thus, although belief in evolution introduced the idea of physical kinship between humans and other, non-contracting social animals, this idea had not at first the power to make available any light on human mental and emotional life. It could not suggest criticism of the arbitrary elements in the Social Contract myth, and did nothing to lessen its influence.

Myths play a crucial role in our imaginative and intellectual life by articulating the patterns that underlie our thought. They are the general background within which all detailed thought develops, and anyone who thinks he is free of them has simply not taken the trouble to become aware of that background. The way in which myths work is often very obscure to us. But, besides their value-implications – which are often very subtle – they also function as summaries of certain selected sets of facts.

A powerful myth, such as the Social Contract story or the tale of Persephone, does, among its other meanings, sum up a crucial range of human experience. Persephone's story crystallizes the way in which good and bad fortune, light and shadow, joy and sorrow are rhythmically linked in human experience. More particularly, it brings home facts about the deep ambivalence of personal relations. The Social Contract story, for its part, lights up facts about the workings of oppressive and non-oppressive government – facts that show the crucial importance of consent for all kinds of co-operation, something which the earlier hierarchical picture of government had thoroughly obscured.

Later, more libertarian forms of the contract myth point beyond this to another range of facts – also important – about the ways in which free and unfree individual choices work. Again, the myth of the Fall of Man draws attention to the undoubted, though depressing, fact that human beings often behave extremely badly, and that even when they don't their motives are often obscure, unreliable and mired in ambivalence. Evil, too, is an important range of facts in the world.

When we attend to the range of facts that any particular myth sums up, we are always strongly led to draw the moral that belongs to that myth. But that range of facts is always highly selective. It is limited by the imaginative vision that lies behind that particular story. This vision can, of course, generate actual lies, which is what makes it plausible to think of the myth itself as a lie. Thus, myths about the inferiority of women, or of particular ethnic groups, have supported themselves by false factual beliefs about these people – beliefs often so bizarre that they would otherwise never have been believed. But even where there are no lies, this selectivity limits the value of myths and calls on us to be cautious in using them. We need always to keep correcting any one of them by countering it with others.

For instance: One regular corrective to the Social Contract pattern has been the use of organic myths, such as the thought that we are members one of another, or that economies and other institutions 'grow', 'develop' and 'mature', or that we share the fate of Antaeus, who drew his strength, like a tree, from his mother the earth. If we are aware of these various myths *as myths*, we can use one to correct another deliberately in this way. If, however, we treat an accepted myth as literal fact – if we simply are not aware of its influence, but swallow the range of propositions it offers as the only one that could reasonably be accepted – then correction becomes very hard. If, for instance, we see 'economic growth' or the growth of empires, simply as a law of nature, a necessity comparable to the growth of organisms, then we cannot think clearly about its actual meaning.

NEGATIVES CAN BE USEFUL

Apart from myths, however, what about the big factual question itself – how *did* human morality actually originate? This of course still remains as a separate issue, one that has a meaning for us now, though it scarcely did for previous ages. But it is one of those historical questions for which we can scarcely expect ever to get a direct answer. Certainly we do now have methods for studying very early history to some extent; we can distinguish moderately reasonable speculations about it from simple fantasy. Physically, we can get some evidence about people's state of health and their food, while their artefacts tell us something about their way of life. But in interpreting their attitudes to these things, we lack that most crucial material – their own words and behaviour.

We can indeed wonder how and when our remote ancestors did actually come to be troubled with a conscience, how they became aware that they could make free choices, how they developed moral concerns to the extent that every human society now has them. But we are unlikely ever to have more than the faintest, most tantalizing indications about this strange process, indications which can mislead us as easily as they can help us. They are misleading not just because they are scanty, but because of our own remoteness. Even if we could somehow listen in at some crucial point and had help with the language – or proto-language – the situation would be so unimaginably strange to us that we would stand little chance of grasping it. So we have here a gap which we have to fill in, like other historical gaps, as best we can from indirect evidence, from what comes before and after, and from careful comparison with other species.

What we now know of that general context does, however, enable us to say some *negative* things with confidence. Some stories cannot be literally true, and among these are both our current guiding myths – not just the Genesis story, but the Social Contract myth as well. The point is not just that there never was a moment of contract. Much more deeply, there was never the need to which that contract would have been an answer. Far from being originally solitary, the earliest human beings were heirs to a long, complex tradition of group life, deep social affection and interdependence, a tradition which dates from many ages before their emergence as a separate species and their famous rise in intelligence.

They share this inheritance with almost every creature on this planet that can be called intelligent. Earthly cleverness is essentially a social phenomenon, an aspect of interaction, closely linked with the power of communication. No creature has evolved as a solitary mathematician. And even if human beings had for some reason wished to withdraw into a more solitary way of life, they could not possibly have done so, because the special developments which raised their level of intelligence demanded of them ever more, not less, co-operation, affection, mutual help and interdependence.

The long, helpless infancy which is needed to develop an intelligent, warm-blooded adult absolutely requires a background of loyal, self-denying, co-operative elders. And speech – notoriously a central aspect of intelligence – would hardly have developed very well among Rousseau's determined solitaries who seldom met and cared nothing for each other's opinions. Neither could any of the cultural activities through which human intelligence largely shows itself have done so.

People, in fact, have never been any less social than they are now. In this context, it seems reasonable to see their capacity for free choice, too, not so much as a private, individualistic rebellion against their social nature, but more as itself a social gift, finding its function in a social context. Normally, we choose together. We help each other to choose, and we make our choices for others as well as for ourselves.

HOW LARGE IS MORALITY?

It is interesting to see how the factual and moral elements are intertwined here. The knowledge that we now have about the social life of early humans and of other species does not, of course, make it impossible for someone to take up the moral position that free, independent choice is the sole and supreme human value. But it does remove the imaginative picture which made that judgement look plausible – the picture of isolated human beings as prospering in an original asocial condition.

If it had been true that they could prosper like that, then they would indeed be a kind of being for whom morality might be primarily an external system of restraints – at best, a set of traffic rules to secure comfortable survival. But morality in all known human history has had a much wider function than that. It has been, among other things, a panorama of ideals, a way of developing the feelings in a particular direction, a set of arts for visualizing better kinds of life, for working together on the understanding of human destiny. And this is surely just what so sociable a species might have been expected to make of it.

Morality is not, then, just rules. All the same, in asking about its history, there is something to be said for starting by attending to the element that makes its development most surprising, which is indeed its restrictive function, its power of imposing rules on desire. How this restrictiveness came to be accepted at all does seem to be the first question that arises. Darwin was right, then, to start from 'the imperious word *ought*'. And this is, of course, the question that the Social Contract myth is especially designed to answer.

11

THE STRENGTH OF INDIVIDUALISM

HIDDEN COMPLICATIONS IN EGOISM

As we have noticed, the Social Contract myth still thrives today in spite of its obvious weaknesses. It is astonishing how easily, in asking about origins, we slip into accepting its assumptions and using its language. Our question constantly takes the Hobbesian form, 'how did an original society of egoists ever come to find itself lumbered with rules that imposed consideration for others?' Thus J. Thibaut and H. Kelley flatly declared in their *Social Psychology of Groups* that 'Every individual voluntarily enters and stays in any relationship only as long as it is adequately satisfactory in terms of rewards and costs.'[1]

And even Thomas Nagel, the most determined and effective resister of this approach among present-day philosophers, writes of

> the central problem of ethics; how the lives, interests, and welfare of others make claims on us and how these claims are to be reconciled with the aim of living our own lives.[2]

But might not most cultures suppose that central problem to be rather that of Plato's *Republic*, 'how we ought to live' in the sense of 'what ought human society to be like?'[3] – a problem internal to ethics itself, rather than one about the possibility of starting on it?

The Hobbesian approach has been built deeply into our thinking because it was the form in which our precious concepts of political freedom and autonomy were developed during the Enlightenment, and it has always been used to express them. The whole idea of what an individual is was reshaped to fit that form, to turn him (though not usually her) into a simple, standardized contracting party, a political unit emancipated from family and friends. This

121

abstraction has great uses in resisting large-scale oppression, but it can scarcely be applied to personal life.[4] In order to 'be ourselves' in the private sphere, we need an appropriate context. Even exalters of solitude like Nietzsche look forward to a better age when they might find suitable companions. Except for the odd hermit, we see actual, thoroughgoing isolation as a kind of death.

The very simplicity of egoistic theory makes it unusable in accounting for most of the actual complexities of life. Perhaps indeed a society of consistently prudent egoists might, if it ever existed, build institutions for mutual insurance quite like those found in actual human societies. And these careful egoists would certainly avoid many of the foolish atrocities that human beings commit. But this cannot mean that morality, as it actually exists anywhere, arises only from this calculating self-interest. People are nowhere near prudent and calculating enough to be like this. Even when they do calculate, they often aim outward, at changing things in the outside world, without thought of how this will eventually affect themselves. This is as true, and as obvious, in the case of adults determinedly seeking revenge as it is in that of children building sand-castles or rescuers impulsively diving in to save the drowning.

Faced with these deplorable tendencies, egoist theorists tend to concede that these things do happen, but are irrational. We all *ought*, they say, to be consistent egoists, even though in fact we are not, because rationality just is consistent self-interest. Indeed, economists and games-theorists sometimes use the word 'rational' with this strangely simple meaning – a piece of unthinking Hobbism which has drifted into technical usage because it makes calculation easier. But the word *ought* is still outstanding here. Officially, egoism exists to explain the force of that word. If it uses the word in its own reasoning it becomes circular.

There is, too, a whole central class of cases where this idea of pursuing one's own advantage does not help at all. How can we decide between various advantages, and between various evils? How, for instance, can we direct ourselves when our life already has what are supposed to be sufficient advantages but it still lacks meaning, when we wonder if it would be better to live differently, when we consider becoming painters or monks or going on an Arctic expedition? What is the place of *ideals* in human life? And how are conflicts of ideals to be arbitrated? Merely bargaining with other people for recognized advantages is no help here.

'OUGHT'?

Contract thinking also notoriously sticks at the well-known question of whether and why we ought to obey the contract. Strictly egoist theory says that there is no ought about it; words like *ought* are just indirect expressions of fear and prudence. But of course freeloaders can often do better for themselves by dodging contracts, and with this use of the words they surely 'ought' always to do so. Nietzsche ingeniously explored the possibilities of regarding morals in this way. He often put with great force the case for a kind of ethical egoism – egoism as a missionary enterprise rather than as a mere clarification of existing fact. But again, he was recommending new ways in which people *ought* to view morality, rather than trying to understand the way in which they actually do view it. So we have to ask again, what is the authority behind his reforming propaganda?

This question is very interesting because Nietzsche himself was deeply concerned with problems about conflicting ideals. He hated complacent, insensitive bourgeois life and thought it ought to be abandoned. But this demand again raises the puzzle about the kind of authority invoked by words like *ought*.

MORAL AND FACTUAL INDIVIDUALISM

In spite of all these difficulties, however, extreme individualism is still very persuasive today, which is why I think we still need to attend to it. Many of the objections just raised to it are old, and I do not think they have been answered, yet they have become strangely hard to hear. Individualism draws remarkable power from two quite different kinds of consideration, one moral, the other factual. As we noted at the start, on this topic these two kinds of consideration are always hard to separate, yet we badly need to distinguish them, and also to trace their connection. At some level our history, our values, and our fundamental nature are indeed linked. This link cannot possibly be a simple one, but any origin-myth that makes it look simple has great appeal.

(1) There is the moral case for isolating people from each other. This rests on exalting a particular group of virtues – notably independence, courage and honesty – and willingly sacrificing all other human values so as to cultivate them. That is individualism as a moral position. It came forward during the eighteenth century as part of Romanticism, and has served as a powerful banner against

both political and domestic oppression. In theory it is quite a possible attitude, and it has been strongly defended by moralists such as Nietzsche and Sartre, disturbed about excesses in the corporate direction and anxious to set the moral pendulum off on one of its endless oscillations.

But these highly rhetorical writers have never interested themselves in explaining how life as a whole – as opposed to life in prison, or a passing adolescent revolt – could be carried on with this very narrow, and essentially negative, set of ideals. Their position can of course always be further refined. But it is unavoidably an extreme one, a stark, fanatical moral outpost for which very serious arguments would always be needed. It cannot make its way – as it largely has up to now – just by making people feel ashamed of their natural dependence on others. It is bound on its own principles to let them judge freely whether they find good reasons to adopt it.

(2) But extreme individualism has also been supported in a quite different way, on supposedly scientific grounds, as a factual discovery. It is treated as a piece of information about how human beings are actually constituted. Today, the most usual form for this argument is still the Social Darwinist idea, which we mentioned in Chapter 1, that evolution proceeds, for all species, by the 'survival of the fittest' in unmitigated cut-throat competition between individuals. That process is held to have shaped them into isolated social atoms, and to be the only mechanism by which they can survive.

Herbert Spencer, who shaped this view, did indeed say that, as civilization progressed, egoism would gradually give way to altruism, so that the social war of all against all would be followed by peace. But he was sure that the time for this change had not yet come. Though individuals might already sometimes experience altruistic motives, it would be most dangerous for them to indulge them in action on any significant scale, because this would hinder the process of natural selection on which progress depended. In particular, there must be absolutely no organized charity to the poor, who were unfit and should be eliminated:

> The whole effort of nature is to get rid of such, to clear the world of them, and to make room for better . . . If they are sufficiently complete to live, they *do* live. If they are not sufficiently complete to live, they die, and it is best that they should die.[5]

This was evidently one of many ways in which nineteenth-century observers who were naturally disposed to humane feeling armed themselves against the sense of horror produced by the visible and spreading misery inflicted by the Industrial Revolution. Since there was no obvious means of relieving it, the best escape seemed to lie in seeing it as inevitable and transient, as merely a stage towards something better. No scruples arose about this policy because it guaranteed a perfect future. By this simple means, said Spencer, 'the ultimate development of the ideal man is logically certain . . . Progress therefore is not an accident but a necessity. Instead of civilization being artificial, it is a part of nature.'[6] In Spencerism, exactly as in Marxism, the prospect of an eventual golden age acted as a blanket justification for objectionable behaviour in the present. Morally, these two systems were, and have remained, extremely close. The chief difference at present is that only one of them is widely seen to have been discredited.

Serious biologists have always disowned Spencer's picture of evolution, which was originally produced quite outside the context of science, and was developed, largely in the USA, as a justification for free enterprise in commerce. Spencer himself was indeed convinced of his scientific correctness. 'My ultimate purpose', he wrote, '. . . has been that of finding for the principles of right and wrong in conduct at large, a scientific basis.'[7] But his story does not even attempt to relate his conclusions to the evidence. It could not possibly justify his wild predictions and moral extrapolations. It is, in fact, just one more expression of romantic individualism. Yet it has managed to pervade the general culture to an astonishing extent, and has lately penetrated academic thought once more as an unacknowledged element in Sociobiology.

Despite its scientific feebleness, however, the Social Darwinist account of evolution is often seen as resting so directly on evidence as to be a straightforward factual history, unlike all earlier stories about origins. Many Victorian readers, already soaked in free-enterprise economics, instantly read it into *The Origin of Species*. Spencer's loose formula 'the survival of the fittest' expressed it to the general satisfaction. Accordingly, the supposed facts of evolution were seen as factual evidence for egoism, replacing that direct, informal experience of human affairs by which Contract theorists had previously supported it.

The story of evolution, conceived in this way, thus became a very powerful supplementary myth, lending quite new support to individualism. Because it deals largely with vast remote, prehistoric affairs, it cannot be refuted by everyday experience, as Hobbesian egoism could. Accordingly, 'evolution' conceived in this way is still widely seen as literal, scientific fact. It carries an air of up-to-date authority which Contract thinking – obviously metaphorical and damaged by centuries of criticism – cannot so easily claim.

Contract thinking, however, has of course not vanished. Today, the individualist imagination shifts constantly between these two pictures, retreating to one when the other is attacked, instead of seriously confronting the faults of either. Since the Social Darwinist picture is unashamedly naturalistic and Contract thought is officially anti-naturalist, this habit can produce astonishing confusions. But these are largely dealt with by compartmentalizing. Supposedly scientific considerations are kept quite separate from moral ones in a way that is often supposed to be dictated by the 'fact–value gap'.

PUTTING COMPETITION
IN ITS PLACE

In the crude form just cited, this Social Darwinist myth obviously is not scientific fact. It contains at least as much emotive symbolism from current ideologies and as much propaganda for limited, contemporary social ideals as does the Social Contract story. But because it is so deeply entrenched in our background, a word more should perhaps be said here about what makes it so remote from current science.

The central trouble is its fantasy-ridden, over-dramatized notion of competition as deliberate, conscious opposition. Any two organisms that both need something they cannot both get are, in a broad sense, competing, and this is all that is needed for natural selection. But they are not acting competitively unless they both know this and respond by deliberately trying to defeat each other. Since the overwhelming majority of organisms are plants, bacteria, etc. which are not even conscious, the very possibility of deliberate, hostile competition is an extremely rare thing in nature.

But the point goes deeper. Heraclitus was wrong. Though conflict and opposition are extremely important elements in every sort of life, it is absurd to suggest that, even at the unconscious level, they could be its central feature. Where there is no cohesion, things simply fly apart and are heard of no more. Life processes, by contrast, depend on an immense background of harmonious co-operation that builds up the system within which the much rarer, though still important, phenomenon of competition becomes possible. In an ecosystem, plants normally exist in interdependence both with each other and with the animals that eat them, and those animals depend both on one another and on their predators. Even at the chemical level, there is a tendency to form bonds and to move towards greater complexity. As Prigogine and Stengers explain;

> We now know that, far from equilibrium, new types of structures may originate spontaneously ... We begin to see how, starting from chemistry, we may build complex structures, complex forms, some of which may have been the precursors of life ... These far-from-equilibrium phenomena illustrate an essential and unexpected property of matter; physics may henceforth describe structures as adapted to outside conditions ... To use somewhat anthropomorphic language, in equilibrium matter is 'blind', but in far-from-equilibrium conditions it begins to be able to perceive, to 'take into account' in its way of functioning differences in the external world (such as weak electrical fields) ... From this perspective life no longer appears to oppose the 'normal' laws of physics.[8]

If there had really been a natural 'war of all against all', a primary urge of all entities towards mutual destruction, the universe could never have taken ordered shape in the first place. It is not surprising, then, that conscious life, arising out of such a background, acts in fact in a way that is much more often co-operative than competitive. And when we come shortly to consider social creatures, we see clearly that co-operative motives supply the main structure of their behaviour.

12

THE RETREAT FROM
THE NATURAL WORLD

———— •✦• ————

EVOLUTION REMAINS
INDIGESTIBLE

ARE WE making progress with our central project of finding some intelligible relation between our evolutionary and our moral thinking? So far, we have been noticing some of the factors which have made this quest so hard. We have noted how, as soon as theories of evolution appeared, attitudes to the problem polarized and how that polarization was expressed in a variety of compelling myths. The myths have offered us only a choice between two package-deals. They promise us *either* a unified theoretical understanding of all nature, including humans, linked with an unrealistic, over-simple view of many things including morality, *or* a more sensitive and realistic conception of human morality, but one which shows it as unintelligibly cut off from everything else in the world.

No doubt in principle we could reject this choice and look elsewhere. But feuds have herded people strongly into the two opposing camps. Disputes – often politically loaded – have raged on various important issues about the status of both religion and science. The questions they have raised about human nature naturally became linked to the evolutionary issue and deepened the division. It has remained extremely hard to break out of it. The price of this feuding has been heavy. In theory, the reductive approach should have lessened the sense of human isolation from nature, by crediting people with the same basic motivational structure as other species. And some of its exponents, especially recent ones such as Desmond Morris, have indeed seen it as deepening our sense of kinship with them in this way.

But for a long time the approach was used entirely for other purposes. It served above all for propaganda within human affairs – to justify, for humans, the political liberty of individuals against tradition and against government, especially in commerce. Until the last few decades, it never really procured any respect for the green environment. And the idea of studying the motivation of other species in order to get a fuller understanding of this kinship never formed part of the reductive project. Most Social Darwinists would have thought it irrelevant. Spencer and his followers supposed, like Huxley, that they fully understood this motivation already. Thus the aspects of nature with which those followers were encouraged to feel kinship were narrow stereotypes, abstractions drawn from a traditional myth and picked to justify a special moral and political programme.

Social Darwinism, then, could not, in spite of its bold, reductive tone, resolve the frightening anomalies attending the new ideas about human origins. It could not smooth that grinding of the moral gears which arose from trying to accept the fact of human evolutionary history. Instead, its exponents, who were committed anyway to debunking certain moral pretensions, have often enjoyed emphasizing the clashes that shock their opponents. They seem usually to have thought that these clashes were just superficial irritants, salutary deflations of human self-esteem, mild resentments that would only trouble a kind of conceit which in any case deserved to be chastened. That conceit is certainly present, and it does indeed need chastening. But it is only half the story.

The meaning of kinship between humans and other animals depends on how those animals are conceived. And, as our imagery still shows, even today, at a deep imaginative level, people still tend to see animals as symbols of odious, anti-human qualities – wolf, pig, dog, cow, raven, rat, toad, jackal, snake – the list is endless. Nor are the images of our nearest relatives, the other primates, much better.[1] It is not surprising, then, that people who think that they are being asked to accept kinship with odious qualities resist the idea. And though animal symbolism does extend into milder areas where it chiefly conveys vigour or freshness or innocent simplicity (lion, lamb, eagle, dove), these meanings are too thin, too abstract and heraldic, to do the work which is needed from a useful sense of kinship. It is not possible to feel related to an abstraction, even a benign one. And since sinister abstractions hugely outnumber the benign ones, people still find it quite natural to think of their more disreputable motives as their 'animal nature'.

TWO NATURES?

Are there, however, really two distinct natures within us? The idea of animality as a foreign principle inside us, alien to all admirable human qualities, is an old one, often used to dramatize psychological conflicts as raging between the soul and 'the beast within'. In Platonic and early Christian thought, this pattern was notoriously central. It might have been expected to die down as Christianity gradually allowed itself to feel more at home in the world, and still more so as its influence declined in any case. Yet this idea is still strong in a surprising number of people who are not religious at all. The pride which used to focus on the soul now centres instead on the will and the intelligence, and it often regards natural human feelings, as well as the body, as alien determinants.

In spite of its overwhelming difficulties, the idea of mind as essentially separate and potentially opposed to the body still seems to be used as a background framework for certain topics, notably for free will, for artificial intelligence, and for our thought about other animals. While this idea has prevailed, only the two stereotyped approaches to our evolutionary status already noted seemed open. People could either take a depressed, reductive view of humans as 'no better than the other animals', or a purely other-worldly view of them as spirits, or pure choosers, or culturally-determined entities of some kind, possibly programmes, inserted somehow during the evolutionary process into bodies to which they bore no real relation.

Hence, then, come the two simple ideas that are now current about the origin of ethics. On the Social Contract pattern all animate beings equally are egoists; human beings are distinctive only in their intelligence, in being the first *enlightened* egoists. On the dualist or separatist view, by contrast, the insertion of human minds introduced, at a stroke, not just intelligence but also a vast range of new, distinctively human motivation, much of it altruistic.

Today, even non-religious thinkers often show an intense exaltation of human capacities which treats them as something different in kind from those of all other animals, to an extent which seems to demand a different, non-terrestrial source. Science-fiction accounts of a derivation from some distant planet are occasionally invoked with apparent seriousness to meet this supposed need, and some quite eminent scientists, such as Francis Crick and Fred Hoyle, have made moves to try to give them a scientific backing. It is, however hard to see how an origin somewhere else in the universe would resolve the problem of continuity between mind and matter.

ETHOLOGY AND REDUCTION

We ought surely, however, now to be able to avoid both these bad alternatives by taking a more realistic, less mythical view of non-human animals. The rich, complex nature of social life among many birds and mammals is common knowledge today. People indeed have long known something about it, though they ignored that knowledge when they were using animals as incarnations of evil. Thus, two centuries ago Kant wrote, 'The more we come in contact with animals the more we love them, for we see how great is their care for their young It is then difficult for us to be cruel in thought even to a wolf.'[2]

By now, however, we know from careful, unsentimental investigation that social traits like parental care, co-operative foraging and reciprocal kindness show that such creatures are not crude, exclusive egoists. They have evolved the strong and special motivations needed to form and maintain a simple society. Mutual grooming, mutual removal of parasites and mutual protection, traits common among social mammals and birds, cannot have been produced by prudent calculation, because the creatures in question are not capable of calculation on this scale. Nor are these habits a deceptive cover for some other motive, because animals are not skilled full-time hypocrites. Social creatures, including all our primate relatives, did not build their societies by plotting their way out of an original war of all against all. What makes them able to live together, and sometimes to co-operate in remarkable tasks of hunting, building, joint protection or the like, has to be their natural disposition to love and trust one another.

The acceptance of this continuity has, I think, been hindered by the unlucky fact that some popularizers of ethology have themselves taken a sharply reductive tone, a tone still calculated to block any perception of continuity with human life. The pioneers of modern ethology, such as Tinbergen, Lorenz and Julian Huxley, did not do this. Nor do those today who have most closely followed their methods. For instance, a remarkable group of primatologists – Jane Goodall, Dian Fossey, Birute Galdikas,[3] Shirley Strum, Frans de Waal and others – has treated its subjects with a seriousness and respect that allows comparison with human life to proceed undisturbed. Against that background, both differences and likenesses can be carefully noted. The sociobiologists, however, tend not to do this, and Desmond Morris has not always done it.

For writers like these, reduction is often a weapon in a wider campaign. Many of them see the continuity between human life and that of other animals as merely one stage among several that are to reveal it as essentially just a physical process with nothing remarkable about it. Their concern is exactly to stress that unremarkableness, to flatten down any notion of the distinctiveness, not just of human life, but of all life. As Peter Atkins has put it, 'Inanimate things are innately simple. That is one more step along the path to the view that animate things, being inanimate, are innately simple too.'[4] Everything, in fact, is a machine really. With this kind of general thesis in mind, these theorists naturally use concepts that – as we have seen – fit human life badly. Readers, not noticing that they often fit animal life badly too, tend to be confirmed in their view that all comparison between humans and animals was useless in the first place.

THE OCTOPOID ANGLE

It is, however, interesting to ask whether any actual animals exist that can take us nearer to the Social Contract pattern. The best candidate that has so far emerged on our planet seems to be the common octopus. Octopuses are self-rearing; they never meet their parents or their offspring. Born as tiny, free-floating creatures, they usually get eaten early in life, yet if they survive to be adults, they become quite intelligent predators, and they do then communicate with one another by coming out in an impressive variety of coloured stripes and spots. They also show their intelligence by escaping most ingeniously from their tanks when captured, and by playing practical jokes on their captors. It is not known what they converse about among themselves. Their topic may well often be territory, but then again, for all we know, it might be the Social Contract.

This matter should probably be looked into. Instead of that, octopuses currently get so little respect that – owing to their imprudence in not being vertebrates – they did not, until very recently, qualify for any of the protection that British law gives to other laboratory animals, including even fish, and the limited protection that is now offered them does not extend to their cousins the squids. This grading of intelligence below backbones has surely been strange. All the same, the octopuses' situation serves to show how far the Hobbist model is from the most basic conditions of human life. Unlike them we are mammals; we spend all our formative years in close dependence on others. Only a form of sociability that accepts that simple fact is open to us.

REDUCTIVE EVASIONS

As we have seen, however, our culture has busily resisted that fact through a remarkable series of psychological theories designed to dissolve away the evidence that human beings naturally have a direct regard for others. It will be useful to return here to reductions already quoted, so as to grasp their implications more fully. Thus, as we have seen, Hobbes laid down the rules for psychological reduction by legislating that

> No man giveth but with intention of Good to himself, because Gift is voluntary, and of all Voluntary acts, the Object is to every man his own Good.[5]

As time has gone on, his successors have invented a variety of techniques for applying this notion to awkward cases. Thus Freud explained that

> Parental love, which is so moving and at bottom so childish, is nothing but the parents' narcissism born again.[6]

More recent fashion, however, has added a new story, or rather two incompatible new stories:

> Parental love itself is but an evolutionary strategy whereby genes replicate themselves ... We will analyse parental behaviours, the underlying selfishness of our behaviour to others, even our own children.[7]

The word 'selfishness' in this last passage does not have the peculiar technical sense of 'gene-promoting' which it is officially supposed to bear in all sociobiological writings, although it would need to have this sense if it were meant to be in accord with the first sentence. The claim occurs at the start of the book, before this special technical sense has even been mentioned. The last sentence does not refer to genes but is meant literally, putting forward a quite different theory, a psychological reduction of affection to selfishness instead of a metaphysical one to gene-activity. And the author goes on to develop this psychological theme, claiming indeed that parents aim at their own advantage (their inclusive fitness) instead of that of their children.

Here, and very often in sociobiological writings, two quite different forms of reduction tangle uncontrollably. Social motives are explained away not only as being – really – evolutionary strategies but also as being really non-social motives; that is, egoistic ones. The persistent determination to cling to Hobbes's unconvincing position remains extraordinarily strong.

LOVE IS NOT A LIE

This position is, however, no less obviously hollow for other animals than it is for humans. The reality of affectionate bonds among social animals is by now fully documented by ethologists. Their sociability is not just a means to an end. It is not something that can intelligibly be dismissed as some anthropomorphic projection. It becomes evident in countless situations – for instance in the unmistakable misery, often leading to illness, of any social animal, from a horse or a dog to a chimpanzee or a human, if it is kept in isolation. Of course this affection does not mean that these animals love each other unconditionally and all the time, any more than people do. They often ignore each other, often clash, and will indeed in many circumstances compete with and attack one another. But they do all this against a wider background of mutual emotional dependence and friendly acceptance.

Devoted care of the young, often including real self-denial over food, is widespread and is often shared by other family members besides the parents. (It may perhaps be seen as the original matrix of morality). Some creatures, notably elephants, will adopt orphans. Defence of the weak by the strong is common and in many well-attested examples the defenders have paid for it with their lives. Old and helpless birds are sometimes fed. Reciprocal help among friends is often seen. All this is by now not a matter of wish-fulfilment or folklore but of detailed, systematic, well-researched record. And there surely is every reason to accept that in this matter human beings closely resemble all their nearest relatives. Anthropological evidence of detailed parallels can be found in *The Tangled Wing: Biological Constraints on the Human Spirit* by Melvin Konner.[8]

CONCLUSION

In this chapter, we have still been occupied with the difficulties which the thought-patterns of our age raise for us when we try to make sense of our position in terrestrial history. We have noticed how oddly those existing patterns force us to regard our own nature, how they offer us a dualism which is both confused and morally eccentric, not to say sinister. And we have noted how one more strong myth – the imaginative identification of animals with vices and evil forces – has joined with those already mentioned to deepen these difficulties. Having glanced, then, at all these problems, we need now to move on and attempt that enterprise itself. We must ask, in what way, if any, is it possible to conceive of human morality as arising out of the kind of innate social tendencies that are found in other earthly species?

13

HOW FAR DOES SOCIABILITY
TAKE US?

———— •◆• ————

NATURAL MOTIVES AS
RAW MATERIAL

WHAT, THEN, is the moral significance of being naturally sociable? That is to say – If we are willing now to accept that humans do have natural social dispositions like other creatures, how do these dispositions relate to morality? Of course they, alone, do not constitute it. Yet they surely do contribute something essential – conceptually as well as causally – to making it possible.

Do they perhaps supply, in some sense, the raw material of the moral life – the general motivations which lead towards it and give it its rough direction – while still needing the work of intelligent reflection, and especially of speech, to organize it, to contribute its form? This suggestion was sketched out by Darwin, in a remarkable passage which uses central ideas from Aristotle, Hume and Kant[1] – a discussion which, until quite lately, received little attention because versions of the noisy egoist myth were widely accepted as the only possible 'Darwinist' approach to ethics.

On Darwin's suggestion, the relation of the natural social motives to morality would be much like the relation of natural curiosity to mathematics and science, or the relation of natural wonder and admiration to art, or that of natural amusability to jokes. These natural motives do not of themselves create the arts and institutions that channel them. But they provide a certain appropriate motivational force that is necessary to create these channels, and they also determine, sometimes in surprising ways, the direction which that force will take.

The metaphor of water and channels – so dear to Freud – has of course a limited use. Motives, unlike water, are not a force which can be turned in any direction. They have a specific point; they can only produce certain kinds of effect. That is why the information model often suits them better than the hydraulic one. Motives carry messages. But these messages do often have a kind of versatility which makes the water metaphor suitable as well.[2] Many motives have some generality. They can express themselves in a variety of ways (ambition and playfulness are obvious examples). Accordingly, if we are talking of original, natural motivation, it is quite reasonable to think of it as, up to a point, raw material, or as a force, perhaps a wind, that is versatile within limits – limits which will be set by the circumstances, especially by the culture.

What, then, about the suggested parallel between sociability, curiosity, wonder and joking? Here indeed are a set of motives which clearly exist, in a general way, prior to culture, because without them culture itself would not exist. Certainly the particular form which they take anywhere is 'determined by the culture'. Yet this determination must surely be in some degree circular, because particular forms of those motives were needed to give the culture the shape that it has in the first place. Not just anything can become a custom.

The parallel with jokes is a particularly interesting one, because it is a case where it seems plausible that the supporting natural motive might really have been absent. Might there be intelligent aliens that were entirely joke-free, totally mystified by any kind of humour? Indeed, are there occasional human beings in that condition? This does not seem unthinkable. Yet if so, then there is something contingent about our sense of humour – something that we might not have had. And we cannot dismiss this as an isolated fact about humour, because the capacity to be amused is not really an isolated power. It is an aspect of playfulness, which is in turn a very important element both in our curiosity and in our sociability.

THE SPECTRE OF CHANCE

What is true of humour may then perhaps be true in some degree of all our natural motives, including our sociability. In some sense, they all seem to be contingent. Over the ages, our species might have turned out otherwise genetically, and would then have had different basic motives. That possibility is surely what makes people uneasy about the idea that the natural affections play any part in morality. The suspicion that beings differently constituted would see no point in duties arising out of our affections seems to undermine their force.

How damaging is this suspicion? The first thing to be said about it is that we are surely asking about our own duties, not about the duties of Aldebaranians or Daleks or intelligent octopuses. Even where moral disputes arise between different cultures, context is relevant. A medieval samurai really did have different duties from a twentieth-century Californian, even where the same principle (such as parental responsibility) underlay them. And a radical difference of motivation, such as that between different species, would make a still more profound alteration. No creature is simply a 'rational being as such'. We all have our particular natures and particular ranges of choice. Our duties have to lie within that range.

That said, however, we want to know how far we can take our own natural motivation as a guide. Here the point surely is that authority does not belong to any single part of it, but in some sense to the whole. No motive is an infallible moral imperative, not even the persistent ones which have been most often treated in this way, such as tribal loyalty and family affection. The mere fact that a motive occurs persistently, in our own or any other species, does not give it automatic authority or turn it into a moral rule. *What makes rules necessary is the fact that motives clash, and clash in the context of a mental life that badly needs to work as a whole.* Having, apparently, more memory, foresight and imagination than other earthly creatures, we are aware, however dimly, of our lives as wholes, and of the way in which serious conflicts disrupt those wholes. In that context, it is not a contingent fact that chronic clashes strike us as presenting a problem. Nor is it contingent that some ways of resolving them seem more acceptable than others.

NON-REFLECTIVE OPTIONS

We will come back to this worry about contingency shortly. But first it is worth while to ask, could these conflicts between motives be resolved in other ways than by a morality? Obviously, the form of reconciliation that is possible depends considerably on the nature of the motives that are present to clash, and also on the kind of wholeness that suits a particular species. If there are creatures somewhere with purely Hobbesian, self-regarding motives, they must experience quite different clashes. They therefore might evolve ways of reconciling them different from any that we can imagine. This may be true, too, of creatures so unplayful as to be quite immune to humour.

That, however, is a matter of remote speculation. The cases that are clearest to us are those of non-human social animals. Among them, some of these conflicts seem to be largely settled by further second-order natural dispositions – inbuilt tendencies to come off the fence one way or the other. For instance, creatures which do a good deal of ritual fighting have certain innate attitudes which tend to fit their clashes into a dominance framework that limits the scope of aggression. And in migratory creatures – as Darwin noticed – the drive to migrate overwhelms all other motives when its season arrives.

These simple traffic rules can be quite satisfactory for creatures who are not given to much reflection. But beings who reflect much on their own and each others' lives, as we do, cannot use them so easily. They do not see each clash as isolated. They tend to remember them and look for connections between them. They therefore need to arbitrate these conflicts somehow in a way that makes it possible to see their lives as reasonably coherent and continuous. In order to do this, they have to set systematic priorities between different aims, and this means accepting lasting principles or rules.

We cannot know whether we are really alone in doing this. There is no way for us to be sure whether any other earthly creatures reflect enough to face the problem besides ourselves. The fact that they do not talk does not tell us much, since a great deal of our own reflection is non-verbal anyway. Experiencing conflict is not just a matter of reciting verbal mantras expressing contrary views. Many people are too inarticulate to verbalize it outwardly, so there seems little reason to suppose that they do so inwardly either. The views of those highly verbal people, academics, have surely been misleading here. A great deal of our inner conflict is not plainly expressed either in speech or in action. Alien observers, even if they knew our language, would certainly miss much of it. It is surely plausible then that, in order for speech to have arisen at all, a fairly complex, pre-verbal, inner life must already have been present among social creatures generally, and that other such creatures still have such a life – as, presumably, human babies also do.[3]

MEMORY, IMAGINATION AND REMORSE

However, a distinctively human element certainly does emerge in the way conflicts are normally handled. Darwin illustrated it by noting the difference between the human reflective predicament and the situation of parent swallows, which can without hesitation join the migrating flock, deserting the nestlings that they have been devotedly feeding and leaving them to die. As he points out, someone who was blessed or cursed with a much longer memory and a more active imagination could not do this without agonizing conflict, which would surely be expressed in behaviour – for instance, in a tendency to interrupt the migration by sometimes turning back.

This sort of vacillation would make serious trouble. If it was not eliminated, the creatures would need to learn somehow to understand their motives better, prioritizing and controlling them. This would involve moral thinking. Darwin therefore thought it 'exceedingly likely that any animal whatever, endowed with well-marked social instincts, would inevitably acquire a moral sense or conscience, as soon as its intellectual powers had become as well-developed, or anything like as well-developed, as in man.'(72) The power of thought, if it once makes visible the conflicts of motive that all animals have, must generate morality.

If so, what principles would thought use to guide it in its arbitration? Darwin notices a most interesting difference between the two kinds of motive that are often involved on this kind of occasion. An impulse which is violent but temporary, such as migration or fury or panic fear, opposes a habitual feeling which is much weaker at any one time, but is stronger in that it is far more persistent and lies deeper in the character. This second feeling is chronic rather than acute. It is less isolated; it is more thoroughly connected to other characteristic motives. Darwin thought that, although the sudden, violent motives must often prevail, reflection, when it did intervene, would tend to favour rules that would protect the milder but more persistent and pervasive ones. Violating these would lead to much longer and more distressing remorse later on, because they resonate with so many other prevalent motives. And these mild but persistent motives would tend to be those that created and maintained social bonds.

In searching, then, for the special force possessed by 'the imperious word *ought*' (92), he pointed to the clash between the chronic social affections and the acute but transient motives which often oppose them. Intelligent beings would, he concluded, naturally try to produce rules which would protect the priority of the first group. Thus, he said, 'the social instincts – the prime principle of man's moral constitution – with the aid of active intellectual powers and the effects of habit, naturally lead to the Golden Rule, "As ye would that men should do unto you, do ye to them likewise", and this lies at the foundation of morality'.(106)

14

THE USES OF SYMPATHY

ON KNOWING WHERE OTHERS STAND

ARE DARWIN'S arguments, outlined in the preceding chapter, convincing? We cannot test his generalization empirically on other species because we have not communicated well enough with any non-human species that we recognize as sufficiently intelligent. (It might be most helpful, for instance, if we could hear something from the whales, let alone the octopuses . . .) So we must simply compare the cases. How suitable do these traits in other social creatures seem to be to furnish the kind of raw material, the kind of favourable wind, that could develop in this way into something like human morality?

Some objectors rule these sociable traits out of court at once because they occur fitfully, and especially because their incidence is strongly biased in favour of close kin. But this same fitfulness and this same bias towards kin still prevail to a great extent in all human morality too. (We will consider the significance of this more fully in a moment.) Yet the bias towards kin is by no means an absolute barrier. The social attitudes that children acquire to those immediately around them are ones which can in principle be extended, in varying degrees, to others when those others appear, and in human society they constantly are so extended. Bonds formed between people who were originally strangers are common in all cultures and can be of enormous importance.

What makes this extension possible? Should we, on Darwin's suggestion, invoke intelligence to explain it? If so, this is certainly not just intelligence in the sense of abstract calculation. It is not just a matter of working out inductively that strangers are quite like one's family and so can probably be treated in the same way.

What enables us to make friends with strangers is partly, of course, just habituation – a tendency to like things and people better as we get used to them. We share this strong and useful tendency with very many other animals. Beyond that, however, there is also something much rarer and more interesting. There is a power of getting to know them. There is a positive capacity for sympathy, for entering directly into some of their feelings and responding to them. There is an ability to put oneself imaginatively in the place of others and to see how it is with them. (The word *empathy* is now often used for this instead of the older word *sympathy*, especially where nothing specially friendly is in question. The distinction can sometimes be important. But our present point is a general one, covering a wide range of capacities, so both words are relevant.)

In ordinary life, we do not often attend to the fact that we possess this power of sympathy. In fact, we usually only notice it to complain that it is not as strong and reliable as it ought to be. Like many other natural advantages, we tend to take it entirely for granted. But this range of capacities seems to be something that many species do not share. Quite likely, none possesses it as strongly as ourselves. At a simple level, experiments have shown some interesting differences in this respect between chimpanzees and rhesus macaques. The researchers

> trained pairs of macaques to carry out complementary parts of a single task. They then tried to get each macaque to do the other's job. They found that the macaques were clueless. Each seemed to have no mental picture of its partner's role.[1]

Chimpanzees, however, could pick up their partner's behaviour and play its role successfully. The researchers cite this as part of a considerable body of evidence that apes, but not monkeys,

> seem to be aware of what they are doing or are able to do, and aware of how their actions resemble or differ from those of others. They also seem to be able to put themselves in another animal's position, imagining what it does or what it knows.

And this is notoriously something at which people are even better than apes.

Of course this power is incomplete and uncertain in its working. We only grasp a sketchy outline of other people's experience and even there we can be badly mistaken. Nor is it always a benign power. We can use our understanding of other people's feelings to baffle them or to hurt them. But, such as it is, it is a power only possessed by highly social animals, and among those animals in a quite special degree by humans.

Our tradition, which has been obsessed with contrasting thought and feeling, has had difficulty in classifying powers like this which resist that taxonomical division. The emotional and cognitive aspects of sympathy are conceptually inseparable. Emotionally, it calls for an intense, unstoppable interest in what others are feeling and thinking, a constant awareness of them which gives their responses enormous importance and makes their opinion of us, and of many other things, a permanent background of our mental landscape. Thus it is not possible that there should ever have been people, or any social animal at all, who were 'strangers to vanity, deference, esteem and contempt' as Rousseau imagined early humans to have been.[2] Cognitively, this same power of sympathy calls for and makes possible an immense enlargement of the imaginative horizon by including the thoughts of others among the things that we can think about. This mass of material multiplies manifold the paths that suggest themselves for our reasoning. And it makes argument possible.

The intelligence involved here is not, then, as Darwin's words might suggest, exclusively an intellectual affair. It is a development of communication, which is, among animals as opposed to machines, always an emotional as well as an intellectual business. As far as we know, this is the only matrix within which intellectual power has ever developed. Whether there could ever be a different one, whether pure calculating intellect could possibly develop on its own, is a matter of empty speculation.

These natural powers of sympathy are indeed a remarkable part of our inheritance. Without such powers, the Golden Rule could never have been conceived at all, let alone applied. Hume was quite right to stress the importance of sympathy for morals.[3] Unluckily, however, Hume was taking sides in the futile debate that insisted on opposing feeling to reason, so he treated sympathy as entirely a matter of feeling.

IS MORALITY REVERSIBLE?

If, then, these dispositions are indeed not disqualified by their narrowness from serving as material for the development of morality, does Darwin's picture become a convincing one? There is surely great force in its first step – in his suggestion that what makes morality necessary is conflict, that an 'unfallen' harmonious state would not require it but a conflict-ridden one must do so. If this is right, then the idea of 'immoralism' as the proposal to get rid of morality, which Nietzsche and others have sometimes seemed to recommend [4] would involve making everybody somehow conflict-free. Unless that were done, we would still need priority rules, not just because they make society smoother, nor even just to make it possible at all, but also more deeply, to avoid lapsing individually into states of helpless, conflict-torn confusion. And of course Nietzsche himself was occupied for most of his time in proposing new priority rules, not in saying that there shouldn't be any. In some sense, then, this fact of conflict is 'the origin of ethics' and our search need take us no further.

It may, however, seem less clear just which kind of priorities these rules are bound to express. Is Darwin right in expecting them on the whole to favour the social affections, and to validate the Golden Rule? Or is this just a cultural prejudice? Might a morality be found which was the mirror-image of our own, counting our virtues as vices and our vices as virtues, and demanding generally that we should do to others just what we would least want done to ourselves (a suggestion for which also Nietzsche sometimes wished to make room)?

Now it is of course true that cultures vary vastly, and since Darwin's day we have become much more aware of that variation. Yet anthropologists, who did the world a huge service by demonstrating that variability, are now pointing out that it should not be exaggerated.[5] Different human societies have many deep structural elements in common. If they did not, there could be no mutual understanding between them, and indeed anthropology itself could never have got started. Among those elements, the kind of consideration and sympathy for others that is roughly generalized by the Golden Rule plays a central part, though of course its application is usually confined to a special group of people.

If we ask 'could there be a culture without that attitude?' we may find real difficulty in imagining how it would count as a culture at all. Mere mutual terror, driving coexisting egoists to form a social contract, could certainly not produce one. It would never generate the myriad positive activities that go to make up a culture. The common standards, common ideals, common tastes, common priorities that go to build a common morality rest on shared joys and sorrows, and all require active sympathy. Morality needs, not just conflicts, but also the willingness and capacity to look for shared solutions to them. Morality, as much as language, seems to be something that could only occur among naturally sociable beings.

THE MEANING OF
THE QUIETER MOTIVES

There is not, I think, anything over-simple or reductive about Darwin's suggestion that when remorse arises, it will usually be because a quiet, persistent central social motive has been violated at the instance of another which was strong but passing. The point is not the strength, nor just the continuousness, but that what is quietly continuous is central. Its demands at any time may be slight, but over time they are constitutive; they make us what we are. This steady undertone of sociality is so strong, says Darwin, that it not only usually determines the direction of remorse, but will finally be satisfied with nothing less than the Golden Rule.

It is, of course, important to recognize the moral contribution of the immoralists here in balancing this social bias by reminding us of other values. From Plato's Callicles and Thrasymachus/ through Machiavelli and Blake to Nietzsche and Jung, they have pointed out the need to acknowledge what Jung called the shadow, the range of powerful individual motives that were getting neglected – anger, resentment, sexual passion, greed, self-fulfilment. They have stressed the role of these less social motives and increasingly, as civilization has got more massive, the role of sheer impulsiveness itself. But, in spite of Nietzsche's efforts, it is not really possible for these moralists, any more than for more orthodox ones, to claim that these are the only important motives or that they should dominate the system.

Darwin does not seem to be making any such exclusive claim on behalf of the 'social instincts'. What he understands by those 'instincts' is, in fact, not just one set of impulses among others, but a whole structured way of regarding those around us, an approach based on sympathy. It involves imagining them as subjects like oneself, subjects who experience life in the same way and are not of a quite different status. There are echoes here of both Hume and Kant, and Darwin quotes both with understanding. This is not just one more philistine piece of reductivism. It is an attempt to make sense of morality as it actually is, while regarding it as something which has evolved, and to explain the oddities of its actual working from its evolutionary history.

CONTINGENCY REVISITED

Even granted sympathy, however, does this whole structure of co-operative motivation still lack the kind of authority that is needed if it is to play its role in generating morality? Is it, even as a whole, somehow too contingent and arbitrary? This suspicion has been strong among moralists in Kant's tradition, and of course there are lop-sided elements even in the general structure of our natural motives which do give colour to it.

For instance, there really is bias, notably the bias towards kin. This bias is strong among the small hunter-gatherer societies that seem closest to the original human condition and it does not vanish, it does not even become noticeably weaker, with the rise of civilization. It is still fully active in our own culture. Any modern parents who gave no more care and affection to their own children than they did to all others would be seen as monsters. We quite naturally spend our resources freely on meeting even the minor needs of our close families and friends before we begin to consider even the grave needs of outsiders. It strikes us as normal for human parents to spend more on toys for their children than they spend in a year on aid to the destitute. Human society does indeed make some provision for outsiders, but in doing so it starts from the same strong partiality towards kin which shapes animal societies.

This same consideration applies to the bias towards reciprocity. Are all our principles really just prudent bargains? This can look plausible because it is true that, if we were dealing with calculating egoists, the mere returning of benefits to those who had formerly given them might indeed be a bargain, a selfish 'investment', as the scientific articles so glibly call it when reducing animals to this pattern. But again, in all existing human societies this transaction appears also, and often primarily, in quite a different light.

Though reciprocity does have a strong secondary use as insurance for the future, it starts life as a spontaneous response to kindness, as appropriate gratitude arising from friendliness shown in the past, something flowing naturally from the affection that goes with friendliness. Its core model is the affection of child for parent, which clearly does not represent a careful calculation of advantage. And among animals, where calculation for the future is in any case so much weaker, this spontaneous, backward-looking feeling is far more plausible as the primary source than prudence.

Frans de Waal discusses most interestingly the prominent part that reciprocity plays in the life of his chimpanzees. As he points out, this is not an economic matter; it concerns social interactions, not exchanges of goods. 'Whether what is involved is the returning of a favour or the seeking of revenge, the principle requires that social interactions be remembered.'[7] As he says, it seems likely therefore that attempts to derive human justice from economic exchanges are mistaken. The principle is more primitive. The chimpanzees directly resent social bilking. But it is important to notice what distinguishes this resentment from its commercial counterpart, namely, the context of continuing personal relations. When a friend lets us down, we are not just disappointed at not receiving our due; we are also upset by the failure of the friendship. This is why ingratitude is so disturbing. The reason why having a thankless child is sharper than a serpent's tooth is not just that we have sunk a lot of capital in the investment, but that we thought we were loved and find that we are not.

THE PROS AND CONS
OF PARTIALITY

How do these narrowing biases affect morality? It is quite true that they need to be – and to some extent are – systematically corrected by the recognition of wider duties as human thought develops. This widening, however, is surely the contribution of the intelligence, gradually extending the social horizons as it shapes institutions. It is not and cannot be a substitute for the original natural affections themselves.

Those affections are unavoidably somewhat narrow, since in evolution they have served the essential function of making possible strenuous and devoted provision for the young. If all parents had cared as much for every passing infant as they did for their own, this provision would have been impossible. In such a casual, impartial regime, few warm-blooded infants would be likely to survive at all. Personal attachment is an essential condition of reproduction among people, as it is among other mammals and birds, and its narrowness is the price of its depth. That is why, despite all the pains of family life, no society, among all the varieties that exist, has tried the anonymous, impersonal child-rearing system so eagerly outlined by Plato in the *Republic*.[8]

Thus, as the sociobiologists rightly point out, heritable altruistic dispositions are not easily passed on unless they make possible an increase in the survival of the altruist's own kin, who share the gene that gives rise to them. But when that does occur, it becomes possible for such traits to develop and to spread through 'kin selection', in a way that did not seem conceivable on the older, crude model that only considered competition for survival between individuals.

Some degree of partiality is, then, built into our social nature. It shows itself, not just in favouring kin, but more widely in the way we form attachments, or fail to form them, with all the people who are of importance to us. We are not creatures capable of loving everybody equally, nor even of understanding fully why we do love some people more than others. Attempts to turn us into such creatures – such as those of the Stoics – have usually ended by seriously damaging the whole capacity for love, not by distributing it on rational principles over the whole population.

FEELING AND REASON

Is this partiality, then, an arbitrary, contingent feature which disqualifies our emotional nature from contributing anything important to morals? Kant thought so, and many other moralists have shared his suspicion in some degree. Kant was responding to the 'Sturm und Drang' movement of his day, which exalted strong feeling, especially sexual love and local patriotism, as an all-justifying force. And here he was surely right. (We have other equally strong feelings for which no one would make that claim.) But this cannot mean that we have to set up a general polarity between feeling and reason in which reason does all the serious work of morals and feeling makes no contribution at all.

The defects of this polarity can be seen in looking at how reform actually works, for instance about partiality itself. As human societies have developed, very many moralists in varying cultures have objected to local forms of partiality, calling for consideration for people currently neglected. In doing this, they have depended quite as much on feeling as on thought. Thought is indeed needed to work out the principles by which institutions must be changed. But thought could never start this work if sympathy or compassion had not first drawn attention to what was wrong, and if that attention had not roused yet further sympathy.

Indeed, at this point the division between thought and feeling becomes quite obscure. Is a person who begins to wonder what life must be like for the members of some oppressed class, and who grows increasingly disturbed in speculating about this, primarily engaged in thinking or feeling? The two activities are conceptually inseparable.

Certainly this person must be clear-headed and must get the facts right. Indeed, curiosity about those facts will be an important part of the activity. But that curiosity is not enough. Insensitive people would not have noticed the trouble in the first place. And someone with an undeveloped heart – someone who simply doesn't care very much what happens to other people – won't be moved by information about it, and probably won't pursue the enquiry. Kant, when he made morality essentially a matter of reason, took for granted an emotional background which he did not notice. Strong sympathy for other people, and indeed a positive passion for justice, are necessary factors if 'practical reason' is to move mountains in the way he wanted. And they are certainly necessary if we are to see anything wrong with partiality.

WHAT KIND OF CHANCE?

Is it, however, just an arbitrary, contingent fact about us that we have the structure of feelings that we do have? There is a way in which it obviously is contingent, but this way is surely a side-issue. In a sense it is indeed luck – for us – that we belong to a species capable of sympathy, or that such a species exists for us to belong to. But then, are *we* there at all independent of this kind of 'constitutive luck' as Bernard Williams has called it?[9] If we were dogs or octopuses or even Hobbesian egoists from Aldebaran, we would not be ourselves at all. What we need to know is: is this feeling a contingent one for *us* as we are now?

The kind of contingency that is disturbing is not some cosmic lottery, dropping individuals arbitrarily into different species. It is a supposed looseness of the connection within us, between the parts of our own nature. It concerns divisions within the self. And the sort of gulf which Western moral philosophers have seen as splitting off all feeling from our rational nature does seem to be one of those divisions.

What we call our reason naturally tends to represent the feelings which are quiet but chronic, feelings which form the main structure of our everyday motivation. But we have other motives as well, which are often stronger. So in cases of conflict we may well ask – Is it a chance matter, with no real meaning in relation to the rest of our lives, which of our feelings are chronic and which are acute but passing? Or is there a deeper structure, which can help to arbitrate these conflicts?

Such conflicts are not rare. They are not confined to the exciting choices which dramatists write about. We very often wonder vaguely, at a quite everyday level, 'Ought I to mind so much about this and so little about that? Is there something wrong, that I keep

149

coming back to this consideration? Am I drifting in the wrong direction?'

PRIORITIES AND FREEDOM

This kind of reflection about priority among one's various concerns is clearly a point where it is vital to assume free choice. Nobody really thinks that determinism helps here. We cannot dismiss these choices fatalistically, waiting to see what we do. As we have noticed, we cannot leave the decision to a mechanical flow of feeling, assuming that it is already unalterably determined by the causal chain.[10] If we try to stand back as mere spectators in our own lives, our attempt is still just one more choice, one more possible policy adopted from among others, but without the advantage of really choosing.

At the other extreme, however, libertarians like Sartre have claimed that we are wholly flexible, that our freedom offers unlimited new directions, and have called on us to create or invent our own values, instead of taking goals as in some sense given. Freedom then centres on detachment from all existing feelings.

This extreme view of freedom is very obscure. In any given situation, only a certain range of responses makes any sense at all. Mere arbitrariness itself is not valuable. And within the range that does make sense there must also surely be some limits on what is emotionally possible. We do not think of such points of decision in Sartrian terms as vacuums to be filled by creative originality. They seem more like particular stages already reached in the course of a journey. From these places several directions are possible. Some of them may be very drastic, including perhaps turning back altogether and going somewhere else. But these possible paths are limited, and the landscape around us has features – towns, rivers, mountain ranges – which must limit them still further.

That there are some such limits is surely not a tragic deprivation, since even the range of choices that we do have often bewilders us. We could certainly not handle an indefinite range. Among those limiting features, the structure of our own motivation obviously plays an important part. It must at least affect the range of possible options. Heroism impresses us because it is rare; it cannot be infinitely stretchable. People may (for instance) sometimes be able to tolerate abandoning the guiding ambition of their lives, just as they may be able to cross mountain ranges. But the fact that this is hard – not just externally but inwardly, in the way that things are hard which go against one's nature – is an important element in the situation. Coherence within our lives is not just something convenient. It is necessary for making any meaningful choice at all.

Beyond these questions about what is possible to our feelings, there are, however, more interesting ones about the positive role of feeling in guiding us – questions which bear closely on our present topic. If (for instance) I am to abandon my ambition, it must be for something else that, in some clear sense, I want more. And want is surely in some sense a feeling. Does this mean that I am the slave of my feelings and am not free?

Kant thought that it did. It seemed to him that freedom meant, essentially, that thought should come before feeling, take charge of it and direct it. So he tried to reduce the imperative of duty to pure thought. This project is much less popular today than the opposite, emotivist reduction, but it still needs attention, because it is still likely to strike us as the only possible alternative.

Kant was surely right in seeing that conscience was not just an urge or flow of emotion. He was right, too, in diagnosing that it presents itself *as* an imperative – as something impersonal, external to our present mood, not as something that we have just decided to make up in a fit of original inspiration. He concluded that this made duty purely a matter of reason – a logically necessary conclusion from the abstract conception of a world of rational beings. Our thought, then, drew this conclusion in just the way that it would draw a mathematical conclusion, and imposed it on our feelings from without.

It did not strike Kant how deeply feeling has to enter into our thoughts if we are ever to form such concepts in the first place. It is very natural to proceed as he did, becoming aware of feeling only when it opposes the policies we are trying to form, and overlooking its part in generating these policies. But a naturally asocial being would never have started on the conceptual path that Kant recommended in the first place.

The general feeling of respect that we have for other rational beings is not something imposed on our emotions from without by abstract thinking. In the political metaphor that seemed so natural to Kant, it is not imposed in the way a colonial governor might lay down laws for a subject people. It has roots also in the natural structure of feelings that is our social heritage. It is itself one powerful motive among others for attending to these topics in the first place, and for developing Kantian concepts. It does indeed express itself in thoughts, and it needs careful articulation if it is to be properly translated into action. It positively calls for coherent reasoning, emerging sometimes into rules. But this intellectual aspect does not conflict with the emotional one. It complements it.

CONCLUSION – THE CENTRAL STRUCTURE

Where, then, does all this lead? I am suggesting, not just that our natural structure of feeling is an important element in forming our morality, but that there is nothing wrong with its being so. It is not irrelevant. It is not, as Kant thought, a disreputable, arbitrary, irrelevant partner weakening the authority of reason, but an integral contributor to the shaping of moral thought. Conscience simply cannot be fitted into the Enlightenment classificatory system that divides thought radically from feeling. As Bishop Butler put it, rebelling against Procrustes with deliberate paradox,

> It is manifest great part of common language, and of common behaviour over the world, is formed upon supposition of such a moral faculty; whether called conscience, moral reason, moral sense, or divine reason, whether considered as a sentiment of the understanding or as a perception of the heart; or, which seems the truth, as including both.[11]

When, then, we feel uneasy about some moral position, not because it is inconsistent but about the attitude it expresses, we are right to take our uneasiness seriously. We do not, as Kant's approach might dictate, say 'this reluctance cannot matter because it is merely a feeling'. We try to articulate the feeling. We work to express it in thought and to see what it amounts to.

To repeat: I am not suggesting that such feelings are infallible guides. There are no infallible guides. But we are surely right to take them seriously as suggestions or warning signals, and what we take them to indicate is something of great importance. *This is the underlying structure of feeling that shapes our thought.* That structure is itself not infallible either. It is complex, and elements in it may need to give way to other elements. As Butler said, conscience is not a simple oracle, it is a faculty of reflection. We may need to change quite deep notions about what matters in life. But we can only do this in response to another element in that structure, because outside it we cannot go. We cannot jump out of our bodies.

For instance: people brought up in slave-owning communities have in many places gradually come to see that custom as wrong, and have eventually abolished it. This was possible because they had within them other attitudes, other ideals, other perceptions, with which slave-owning conflicted. Those attitudes and ideals were an important part of their structure of feeling, even though they may not have been explicitly acknowledged in the culture.

Does that structure then have to have an innate basis? This is a complex question.[12] But clearly we are not just abstract 'rational beings as such'. It is not a matter of chance what species we belong to, nor, in general, what kinds of things can be important to that species. 'Free creation' of values only makes sense inside those limits, which are not themselves arbitrary. As Philippa Foot pointed out, someone who says that the only thing that matters is not to tread on the lines of the pavement, or to clap one's hands once in every hour, would find that people were unwilling to accept this as an impressive piece of original moral thinking. Despite their novelty, these suggestions do not count as creating values.[13] By contrast, someone who draws attention to kinds of pain and misery that have not been previously noticed and asks that something should be done about them will be understood as saying something relevant to morals, whether it is accepted or not.

This is not unfair discrimination. We know that the first set of suggestions are too remote from the centre of human life, that moving too far from that centre makes standards unintelligible. Unless such moralists can explain their proposals in terms of something that plainly does matter, such as health or self-knowledge or annoying the neighbours for some good reason, it will simply count as an obsession, a 'compulsion' – as a meaningless limitation on freedom, not as an expression of it.

Freedom, then, is not essentially a matter of the unrestrained creation of entirely new values, nor of a total detachment of choice from feeling, nor of the dominance of thought over feeling. So what is it? We will move on to try for some more positive suggestions in Part IV.

Part IV
WHAT KIND OF FREEDOM?

15

ON BEING TERRESTRIAL

VERY NEGATIVE FREEDOM

FREEDOM, LIKE purity and absence, is a negative notion; it always means freedom *from* something. That something varies greatly in different situations and often seems so obvious that it need not be mentioned. If you are in prison or in chains or wholly subject to someone else's will there is no doubt what change you are after. On a political scale, however, doubts do arise. Attempts to break one kind of control often involve accepting another. This easily leads to conflicts. For instance, people whose nationalities form part of an empire often see freedom simply as the absence of that empire. But when it breaks up, nationalism itself often proves oppressive.

At such points the unit of freedom changes. Groups smaller than the original centre of revolt protest against it. In principle, this sub-dividing process can go on until each person rejects all interference by those around them. And, as anarchist theorists have pointed out, there often is indeed something objectionable in that interference. But since total confusion suits nobody, some kind of compromise is usually struck. If one takes the ideal of freedom seriously, the aim then seems to be that people should willingly accept restrictions on relatively minor matters, while remaining free to choose on whatever points they think are most important. This modest hope seems to be what we have in mind when we talk – paradoxically, from an anarchist point of view – of a free society.

Has this pattern any analogy inside each of us? The notion of inner freedom is an old and powerful one, but its meaning has varied, just like that of its political counterpart, according to the various kinds of inner tyrant that have been feared. Moralists from Socrates to Spinoza and Kant pointed above all to the tyranny of the passions, urging us to free ourselves from it by the use of reason. And they linked this advice closely with the need to free the spirit from the tyranny of the body.

This notion has a lot to be said for it in so far as it does call for integration. Moralists were quite right to point out how damaging it is for the whole person to be carried away, helpless and unthinking, by any single motive. That sort of disintegration does seem to be the central evil that the notion of inner freedom exists to prevent. The rationalistic sages, however, did not just call for integration. They also took sides against certain particular motives, notably the affections involved in personal relations and some – but only some – of the tastes concerned with self-fulfilment. (Intellectual enquiry always did unreasonably better than art.) The Stoics were particularly vigorous in building these irrational prejudices into the idea of rationality, and their influence remained very strong with the rationalists of the Enlightenment.

It is not surprising, then, that during the eighteenth century the emphasis changed. Romantic individualism increasingly identified the self with the passions and called upon it to resist tyrannical Reason – a point central to Blake. Moreover, a greater knowledge of history and anthropology brought the tyranny of custom in sight as well, by showing how much customs could vary. This kind of tyranny is a central theme is Mill's *Essay on Liberty*. More narrowly, too, domestic tyranny – the tyranny of the family over adolescent children, particularly sons – became a favourite theme of fiction and suggested a notion of the true self as something more solipsistic still, something so delicate that it would be injured by any relation with other people. This terror of involvement seems to have been what Sartre expressed in the opening remarks quoted in Chapter 10, p. 111. For him 'man', the genuine individual, only begins to exist at the point when he (and essentially it is he) is saved from personal contact by cutting himself off from everybody around him.

Since these shifts all call for different kinds of behaviour, their upshot has partly been mere confusion. But, as we have seen, the general trend has been to shrink the central character involved – the free self – making it steadily more abstract by declaring war on its outlying provinces. It is true that the various theorists involved have often tried to suggest subtler and more positive compromises. Each of them, studied in detail, has more to say. But their great interest in opposition itself has enforced the simpler side of their messages. It has naturally suggested a kind of person who is dedicated chiefly to revolt as such, a person who is perhaps so negative as to oppose every kind of control equally. Among these different threatening kinds of control, however, the notion of control by our animal nature has steadily held its place. Although current moral ideals are officially much less ascetic, much less world-denying than they used to be, notions of human dignity still seem to demand a wider gap between ourselves and other species than current biology can deliver.

BETWEEN A ROCK
AND A HARD PLACE

That clash is central to this book. The account that I have been trying to work out of our freedom, and of the sources of ethics, has been aimed at finding an unsuspected middle way. On the one side, we must surely reject the crude, mechanistic, reductive accounts of motive which have so often accompanied insistence on our animal origin, and the fatalism that goes with them. But I think that we have to reject, just as strongly, the no less unreal vision of an anti-septically isolated human essence, a purely spiritual or intellectual pilot arbitrarily set in a physical vehicle which plays no part in his or her motivation. This vision has not just tended to make the story of our origin as a terrestrial primate species look unbelievable. It has also landed us with a notion of our whole nature as unintelligibly divided.

Of course human morality, like the rest of human culture, cannot be reduced to, or equated with, anything found among other social creatures. But some crucial aspects of its working can be understood if it is seen as growing out of that shared social background. Its peculiarities then fall into place as belonging to the history of this particular species – a history which is itself unique. Such an under-standing of its origin will, of course, still always be incomplete. But it does help towards a better grasp of its nature, by removing the obstacles posed by these two unrealistic stories.

THE SIGNIFICANCE OF ORIGINS

As we have seen, the factual and evaluative aspects of this whole topic are deeply intertwined. Though it is wrong to denigrate or exalt things simply by identifying them with their sources, yet when an aspect of life – such as morality – is, in its present state, somewhat mysterious and puzzling, and when its actual history is obscure, speculations about its origin inevitably do have a real importance in framing our idea of it. Our choice of possible causes for them both reflects and influences our present attitudes. The various myths that I have been discussing do this markedly. And I have suggested that the Social Contract story, which is the really influential myth today, not only still makes us distort the facts, but is also dangerously one-sided morally.

There is real difficulty about trying to correct such myths. It is no good trying to put forward a rival account that is myth-free, purely factual and objective. The academic's dream of pure, sanitized objectivity only leads us to conceal essential material. Every argument on important subjects proceeds from a particular imaginative viewpoint, a whole life-position, a vision of human destiny which always has some moral bearing on the direction of thought. That bearing needs to be expressed as clearly as possible, so that it can be used or criticized by standards distinct from the logical ones that apply merely to the consistency of arguments.

The best way to provide for this is, I think, to make such visions explicit. Of course it is useless trying to invent new myths deliberately to replace the old ones. But it is quite possible, and often useful, to express aspects of one's own vision in imaginative form. One thing that seems badly needed here is to make clear to ourselves just what it is that we so much value about human freedom – what we feel deprived of by doctrines that seem to diminish it. So let us now try out a fable directed to that point.

* * *

CREATION TROUBLE

There was once a creator who wanted to create free beings.

The other creators, it seems, didn't share this ambition, indeed they thought the project was philosophically confused. They were well satisfied with their own worlds. But our creator, whom we will call C, sat down to work it out.

'How will you even start?' asked D, the Doubter, who was watching.

'Well, I know what I won't do,' answered C. 'I won't just give them an empty faculty named Desire, and tell them to invent values and to want what they choose. Unless they want something definite for a start, they won't even be able to start choosing.'

'That's right,' said D.

'So what I think I must do,' C went on, 'is to give them a lot of desires which sometimes conflict, and make them bright enough to see that they have to do something about it.'

'If that means giving them a desire to drink and also a desire not to drink, it sounds like a good way to deadlock everybody,' D objected.

'Yes indeed,' replied C. 'So what I propose, instead, is to give them the sort of quite numerous desires which animals usually have, which don't clash all the time, but are bound to clash sometimes. Then the clashes won't be between *drink* and *don't drink*, but between *drink* and *finish building the house*, or between *drink* and *carry drink to your child*. That won't deadlock them, but they'll have to think about it.'

'Are you going to make their thoughts affect their actions, then?' enquired D, somewhat alarmed.

'Well, I believe I am,' said C. 'You know how, when they don't, creatures are inclined to get lazy, and not to bother with thinking at all? Somebody has pointed out that, "when reason is imparted to favoured creatures on top of their infallible instincts, it only serves them for contemplating the happy disposition of their natures, for admiring it, for enjoying it, and for being grateful to its beneficent cause."[1] And, nice though that is, it somehow doesn't seem to make for a very active intellect. This is actually a thing which has bothered me about a number of worlds.'

'I can see that your lot will be kept busy,' said D thoughtfully. 'They will have all sorts of reasons for making all kinds of choices. No doubt they won't be deadlocked, but can you stop them all having breakdowns from trying to do too many things at once?'

'I mean to leave a lot of simple, straightforward motives in them as ballast,' answered C. 'They will all be fairly similar, for a start. They will be instinctively affectionate and sociable. This will give them some steady, lasting purposes in life. It will tend to keep each of them from being carried away by ideals which mean nothing to others. Then, they can consult one another, which should help choice. And the intelligence which they develop will not be mere abstract calculation. It will be an imaginative activity. Other people's feelings and responses will constantly figure in it.'

'Is that necessary?' asked D. 'Won't they be a sufficiently standard product, so that what is sauce for one of them will always be sauce for the others?'

'No,' said C. 'I thought this would surprise you. Though I'm making them rather alike, I'm not going to standardize them completely, and I'm not even going to integrate each of them completely intellectually. As I see it, freedom demands that individuals should be genuinely different, so that they really do have to think for themselves. I mean to leave them the kind of individual differences which animals usually have, and even the wide temperamental differences produced by sex and age. Each one will find itself not only, in a way, alone among aliens, but also a committed member of a number of different tribes, all of which have other, irremediably different tribes to deal with. I think all this will force them to notice what their own various motives are, and to stand back from them, trying to bring them together. Each of them will have to try to operate as a whole person. And that is surely the point of freedom.'

'How will they resolve their disagreements?' asked D.

'Well you see, this is my real problem,' C replied. 'There are two possible extremes. On the one hand, I could make them so harmonious, underneath it all, that in spite of their differences they would never really be in opposition, because they would feel themselves to be parts of one vast whole. Their swarm or hive would be their real individuality. Or, on the other hand, I could have them feeling so individual that they didn't give a damn for each other, and were happy to declare a war of all against all. The first sort might indeed evolve a real way of thinking, but it would be a corporate one. The second sort couldn't evolve one at all, because they couldn't really communicate; their lives would be far too separate. Though they might bargain negatively for self-protection, they could never think or act freely together.

I want the best of both worlds. I want real individuality to the extent that, if one gets its way and another does not, the second has really lost something. But I want them to understand this situation, and be able to choose it. I'm not satisfied with the simple, harmonious, corporate solution, because I don't think it allows freedom. I think each being needs to face real conflicts, both within itself and with those around it. And in both sorts of conflict, whoever loses must really have lost something.'

'And *how* did you say you were going to make them settle for that?' asked D.

'Well,' replied C hesitantly, 'I thought that the best way of doing difficult things is often the crude, obvious one. It's no good giving them an intellectual proof that they really prefer harmony to getting their own way. That just leads back to the corporate solution. I thought I would simply land them in this mess from the start and let them get used to it. They get born into a group of others who are different from themselves, on whom they are emotionally dependent, but with whom they are bound to disagree. Quarrels arise, and they just have to deal with them.'

'Quarrels?' said D. 'Are you going to make them, not just different, but capable of aggression as well?'

'That's right,' said C. 'For freedom they will need these real disagreements, and that struck me as a good way of providing them.'

'No doubt about that,' replied D. 'Well, it's your world. Has it struck you, by the way, that, with all these problems on their hands, your beings may get the idea that they are creators themselves?'

'Yes indeed,' replied C. 'Will they be wrong? I'm really not sure. But I think it's time I got started. Could you just pass me those compasses?'

*   *   *

FREEDOM NEEDS
PLURALITY OF AIMS

At this point, plainly, we must leave the language of fable, before someone complains that this is a shockingly irresponsible transaction. From now on, we will speak like good children of the Enlightenment. We will say that it was not C or D, but E for evolution that did all this. We will insist, too, that evolution has a small 'e' and is not some kind of force or deity. In order to avoid Enlightenment superstitions as well, we will add that the working of evolution does not prove some kind of dogmatic atheism either, any more than it shows that matter is inert and that the only causal factor at work is a mysterious thing called Chance. If a creator was in fact present, then he, she or it seems to have worked through evolution, largely, though not at all necessarily exclusively, through natural selection. (Darwin never thought that this was the only force involved and nor need we). If no creator was present, then the complex processes involved in evolution did the job on their own.

I am not suggesting, either, that those processes were aimed at producing beings like ourselves. That would be monstrously narrow and anthropocentric, as well as mythical. The point of my fable is to explore the meaning of human freedom, now that we have got it. It aims to point out, by showing how the sort of multiple motivation that is normal for animals is actually needed for freedom, that there is no clash between this human freedom and our evolutionary origin. In fact, this freedom is of just the kind that ought to be expected in evolved beings. Our evolutionary origin not only accords with it, but helps to explain it. *An evolved being is not one made like a machine.*

Unlike machines, which typically have a single, fixed function, evolved organisms have a plurality of aims, held together flexibly in a complex but versatile system. It is only this second, complex arrangement that could make our kind of freedom possible at all.

VAGUENESS IS NOT ENOUGH

What, then, is the most important point about human freedom? As we have already noted in Chapter 8, it is not really unpredictability. Freedom is certainly not an indefiniteness, a vagueness of the will which is sufficient to baffle prediction, a looseness like that in a machine whose screws have not been tightened. This strangely negative idea misses the point altogether. As we have noticed, we do not even necessarily think it objectionable to predict the behaviour of a free person. We do this whenever we think of somebody as reliable. The crucial question is, what sort of data do we predict it from? Is the prediction based on reasons and motives – data that would make sense to the people acting – or merely on causes external to their inner life, facts about the things that surround that life?

It is surely only the unsuitable style of prediction, the thingbound style, that insults freedom. At the point where you can predict my acts directly from my state of health or my social conditioning or what I have just eaten, you can stop treating me as a free being. And if you positively choose to predict them in this way, rather than trying to make sense of my point of view, then you are choosing not to treat me as free. You are deliberately regarding me merely as a thing, that is, as part of the surrounding processes. And that is a crucial change of attitude.

This distinction still holds even though the predictions themselves might be the same in both cases. For instance, one observer might predict that someone's suicide has become less probable from an improvement in the sufferer's health and another from a change in his or her beliefs, two things which might occur together. Both considerations are relevant, but to insist on using either alone is to take up a special kind of attitude to the person involved. Whether that attitude ought to be taken up is a moral question, not a scientific one. As we noticed, when heroic people do exactly the heroic thing that we expect from them, bystanders who understand their principles do not say gloomily, 'how predictable'. They do not suspect automatism. They reasonably think that these agents are *more* free than someone who is unpredictable simply by being weak and disorganized, not less so.

A different set of observers, less sympathetic with those principles, might predict the same actions but on quite different grounds, attributing the resistance to ignorance, or to faults such as obstinacy or pride. This prediction, though it might still mention motives, would be much more like a forecast about whether a particular bridge would break under particular traffic strains. It pays less attention to the agent's viewpoint. These remoter observers would be reasoning causally from past experience. They might well mention motives that do not figure among the reasons that agents give themselves, or might attribute the action directly to physical factors such as disease. Their explanation, in fact, tends to bypass the agents' choice entirely.

This causal kind of explanation is, of course, in itself quite legitimate, indeed it is necessary. Normally we need both kinds, using one as a check on the other. Human beings are indeed things and organisms and parts of the processes around them as well as persons. They are also far too complex to understand themselves fully. Other people's view of them can, therefore, quite properly take in facts about their lives that they do not know. And because life has to go on at a certain pace, these judgements must often be somewhat crude and simple.

What really is unjustifiable is the attempt to get rid of the person-centred pattern, to use the causal, thingbound model as the sole explanation of conduct. The reductive notion that only that model is 'scientific' is simply a way of legitimating that deeply unjustifiable policy. It is this reductive threat that has led to excesses on the other, separatist, side – to a whole series of defensive attempts to split the free self off from the rest of the person and to place it somewhere outside physical causes.

STOP-GAP SOULS

As we have seen, this separatist approach is damaging both morally and metaphysically. Morally, it carries an unreasoned bias towards whatever component of the self is deemed to be free – a bias which becomes more harmful the more narrowly that component is defined. Notions of 'the will' or of 'reason' are never morally neutral. They are always dramatic representations of a particular kind of person, so that 'rationalism' has never been as impartial as its champions suppose.

For Plato, 'reason' was naturally embodied in an aristocratic Greek intellectual highly trained in logic and with a strong sense of civic duty, not in poets, rebels, foreigners or shopkeepers. Aristotle had more room for poetry but thought civic duty less central than Plato and was firmer about finding very little reason in women. Both of them thought reason quite compatible with a profoundly religious attitude. Hobbes, by contrast, thought it demanded atheistic, secular individualism and excluded a sense of civic duty. Hume agreed about the atheism but not about the individualism, and tried to allow the feelings some voice against reason. For Kant, reason was above all a defence against the excesses of personal feeling and the will was simply reason in action. Nietzsche and Sartre, however, exalted the will itself as a distinct force on its own, an expression of pure individuality, something that required a much more deeply individualistic morality.

These differences (which I have exaggerated a little for clearness but have not really distorted) are essentially moral differences – differences between ideals, between preferred ways of life. They have, however, always been entangled with views about personal identity, metaphysical views that have looked somehow more factual. Reason or the will has been seen in separatist style as the real self, a particularly solid entity, somehow distinct from, and more real than, outlying parts of the personality such as the feelings. The Newtonian revolution, however, put this drama in a new and more alarming perspective. Physical science now displayed a new map of existence, a map which was so impressive that it seemed to be metaphysically comprehensive – a complete map of all that existed. This meant that a place needed to be found on it for the real self – a place which would still allow that self to influence action.

This map, however, had not been designed to be used for moral purposes but for purposes of speculative knowledge. It could no more accommodate the real self, as morally conceived, than the political maps in the atlas can accommodate mountain ranges or a map of the world can show details of a town. The attempt to make it do so was a category error, produced by the still more disastrous category error of supposing that any one map could be comprehensive in the first place.

Separatist moralists responded, not by fetching a more suitable map, but by shrinking the free part of the self into a still smaller compass in the hope of fitting it in somewhere. This minimal self probably reaches its limit in Sartre's doctrine of the will, an entity that excludes all the feelings which might seem to have a physical basis. But this self is still not small enough find a place in the causal process, because science describes that process in terms of events, not of actions.

While theories about the physical world still contained large and obvious gaps, proponents of free souls could still hope to insert them there as an extra, quasi-physical causal factor. Thus Descartes thought he had given the soul adequate power by putting it in charge of the pineal gland. This was essentially the same move which theologians made when they rashly inserted a 'god of the gaps' to fill holes in the scientific account of the cosmos. In both cases the move is doomed, not just because later scientific explanations sometimes fill these gaps, but because, even where they do not, the assumption of completeness that underlies modern science demands that they must be expected to do so. In a manner which is mysterious but is certainly encouraging for researchers, this conception of science assumes its own possible completeness.

Accordingly, serious theologians long ago saw what is wrong with the idea of a 'god of the gaps' and abandoned it – a fact which somebody should explain to old-fashioned atheists such as Francis Crick and Richard Dawkins. The Soul of the Gaps, however, still survives and tries to prosper. Efforts are always being made to find loose places for it in the causal sequence. Quantum mechanics, revealing some sort of discontinuity in causal processes, has been ardently welcomed as its home. But it is surely hard to see how an active self can be imagined to act in the world simply by interfering with the movement of protons. One cannot, after all, drive a car by sitting inside the engine and redirecting loose cogs. And even the metaphor of car-driving is, as we have seen, misleadingly separatist for the relation between mind and body.[2] Similarly, some have hailed the wider scepticism about determinism that has been introduced by chaos theory as making room for human freedom. But mere unpredictability is, as we have seen, quite unable to do this.

That is not to say that these changes are not relevant. It is perfectly true that the shift in the whole nature of physics during the twentieth century has made the scientific world-picture much more hospitable to multiple types of explanation than its mechanistic predecessor was. And these two innovations have indeed been central parts of this change. At the highest level, this new conception of science is far more aware of philosophical difficulties, far less dogmatic, less over-confident and less imperialistic than the former one. It does therefore leave more room to relate science to other forms of thinking. As Heisenberg put it:

Whenever we proceed from the known into the unknown we may hope to understand, but we may have to learn at the same time a new meaning of the word 'understanding'. We know that any understanding must be based finally upon the natural language because it is only there that we can be certain to touch reality, and hence we must be sceptical about any scepticism with regard to this natural language and its essential concepts. Therefore, we may use these concepts as they have been used at all times. In this way modern physics has perhaps opened the door to a wider outlook on the relation between the human mind and reality.[3]

Since good science is now prepared to relate itself to conceptual schemes outside it in this way, there is surely less reason – not more – for trying now to insert the subjective viewpoint somewhere within the systems of physical science itself. The idea that particle physics contains a suitable soul-shaped cranny capable of accommodating agency and subjectivity seems quite to underestimate the size and awkwardness of these topics.

THE QUEST FOR WHOLENESS

The whole approach centring on predictability has been thoroughly explored, and it has never given much satisfaction. I am suggesting, therefore, that we move right away from it and concentrate on the idea just mentioned – that human freedom centres on being a creature able, in some degree, to act as a whole in dealing with its conflicting desires. This may sound odd, because freedom sounds like an advantage, and having conflicting desires certainly does not. But it is not a new thought that freedom has a cost. And the conflicting desires themselves are of course not the whole story. They must belong to a being which in some way owns both of them, is aware of both, and can therefore make some attempt to reconcile them.

The more clearly that being is aware of the clash, and the more it can, on occasion, distance itself from any of its impulses, feeling itself to be a whole that contains them all, the freer it becomes. This distancing does not mean taking flight to an entity immune from the conflict. Only misguided attempts at self-control are made in that way. The endeavour must be to act *as a whole*, rather than as a peculiar, isolated component coming in to control the rest of the person. Though it is only an endeavour – though the wholeness is certainly not given ready-made and can never be fully achieved, yet the integrative struggle to heal conflicts and to reach towards this wholeness is surely the core of what we mean by human freedom.

16

WHAT KIND OF BEINGS
ARE FREE?

ANGELIC ASPIRATIONS

CONCEIVING FREEDOM in the way I have suggested in Chapter 15 seems to me to be not only natural but unavoidable if inner conflict really has the importance in our lives that I have been suggesting. Is it, however, true that this objectionable thing, conflict, is needed for freedom? To consider this, we need to contrast our conflict-ridden state with some possible alternatives. The first obvious contrast is with the imagined condition of a paradisal being which never experiences any clashes at all, because its impulses are always and inevitably in harmony with each other and with what its situation demands. This is the supposed situation of God, approached, but not quite reached, by that of angels and of humanity before the Fall.

That condition seems plainly better than our present confused state, and yet, paradoxically, it is something that we surely could not accept if we were offered it. This inner simplicity would involve stopping being what we are to such an extent that the change becomes quite inconceivable. Even those of us who are most cross with their destiny for making them as they are, do not usually complain of this feature of the work. They want to eat of the fruit of the tree of the knowledge of good and evil. And you can hardly do that if evil means nothing to you, if it has no hold on you at all.

This, at least, seems to be the human situation. The point has been made imaginatively in many stories, of which *Brave New World* is still one of the best. Science-fiction sometimes continues to illustrate the point by displaying people who are offered operations that will entirely free them from conflict. There is never much doubt how the prospective patient will react.

No doubt if we leave the human situation and speak of God and angels, things do become different. But then, not surprisingly, we have great difficulty anyway in imagining their condition, and in finding a language to describe it. As theologians have pointed out, the word *free* is only one of many terms of praise that are hard to apply intelligibly to God. Language, after all, has been developed to apply to our own species. And here, inner freedom does seem to presuppose conflict. Paradisal beings would be free only in the weaker sense that they are doing what they want. As Hume pointed out,[1] this is an enormously important advantage, and one which other animals seem to enjoy. But it does not measure up to what we usually mean by human free will, which is thought of as something distinguishing humans from other animals.

OTHER SPECIES?

We had better look at this point next. What about the motivation of non-human animals? Here the clashes of motive are often visibly present. Often indeed they appear in hesitant or confused behaviour very like what we display ourselves. This is not just a vague, anthropomorphic observation. Ethologists have carefully analysed and documented the mixed behaviour that arises in these mixed situations. Here is one out of many such analyses, sorting out the contributions of various motives in the preliminaries to a fight:

> In nearly all teleosts [fishes], the fight is preceded by threatening movements which, as we have already described, always arise from the conflict between aggression and the escape drive . . . In many perch-like fishes . . . each of the two adversaries swims straight at the other, preparing but not quite daring to ram home a warning thrust. Their bodies tense and twisted like S-shaped springs, the opponents swim slowly towards each other and come to a standstill head to head . . . Depending on the conflict situation from which frontal threatening arises, they do so not resolutely but rather hesitantly . . . [These hesitant movements, however, serve to display their respective strength and size, so that] Such a prolonged introduction fulfils an extremely important function, in that it enables the weaker rival to withdraw in time from a hopeless contest.[2]

To common sense, this proceeding looks very like human conflict behaviour. Is this apparent likeness deceptive, disguising processes that are really quite different? Much academic thought has held that there is this deep difference. Animals, on this view, cannot really experience a clash of motives because they are not self-aware. Indeed, from this angle it can scarcely be said that a clash even takes place at all. Even if the creature is in some sense conscious, it is thought not to be the kind of conscious being which can own both desires. In this sense it is held not to be 'self-conscious', and therefore not in conflict. One desire simply replaces another, like successive waves on a beach. There is no fixed scene within which they can figure as competing. There can be no drama, because there is never more than one character on stage at a time.[3]

This seems a strangely exaggerated story, one which would surely not have occurred to anybody merely from watching animals attentively, as in the above example, nor indeed from any other reason except the assumption of human uniqueness. Conflict behaviour where the motives are both present together is in fact quite common in animals. And the more advanced the animals are, the more sophisticated the conflict grows. It can be quite prolonged, involving much hesitation, oscillation, intention movements and displacement activity. It then tends to cause visible distress with every evidence of stress and strain, including gastric ulcers. It really is not at all like the succession of waves on a beach.

AMBIVALENT APES

For example: Adrian Desmond, in a fascinating discussion of the attitudes of chimpanzees to various kinds of bloodshed, describes their muddled, tense and ambivalent behaviour on several occasions when they have killed one of their own species. In this instance, well-known individuals from the Gombe group had met a strange chimpanzee female carrying a baby, which they at once seized and killed as they might have done a pig or a monkey:

Humphrey was beating its head against a branch; then he started eating its thigh muscles and the poor infant went limp. Mike was allowed to tear off a foot. But now confusion seems to have overcome attendant apes. They watched intrigued, but none begged a portion. They did however inspect the carcass, and Humphrey too began poking and sniffing rather than eating it. He even groomed it, then dropped it and walked away (*prey* is devoured by the group with not so much as a scrap wasted). Others retrieved the small corpse, only to play, examine or groom it, often giving it the respect accorded a dead community member. The carcass changed hands six times and, although battered beyond recognition, very little had been eaten.[4]

If, following Darwin's suggestion, we imagine conflict situations of this kind recurring, as they would be bound to, repeatedly over time in a pre-human species, while the power of the creatures' intelligence and the scope of their memory was gradually increasing, it is surely plausible that they would begin to feel a need to find some way of drawing clearer, firmer lines about what behaviour was appropriate, both at the species-barrier and within the species. Custom would begin to dictate more firmly what could and could not be done to whom.

For this to happen, however, the rise in what we call intelligence must, as we have noticed, be more than just a cognitive sharpening, a growth in the power of making abstract calculations. The full idea of intelligence always involves a *directed* curiosity, a grasp of what is important and a peculiar interest in it. To produce useful thought on topics of this kind, there must be a special concentration of attention on it which makes this particular confusion seem intolerable. A more active power of sympathy is surely a necessary aspect of this increasing awareness.

In discussing these disturbing cases of chimpanzees' violence against their own species, Adrian Desmond draws attention to an important emotional difference between their response and that of human beings. Chimps do not punish on behalf of their community. This does not mean that retributive notions are entirely absent. As Frans de Waal shows, the apes can visit personal resentment afterwards on an individual who has failed to back them up during a contest.[5] Interestingly, too, they do seem to expect that their leader shall intervene in contests to protect the weaker combatant,[6] and if they are present they do quite often defend those currently being attacked. But they seem to display no indignation against aggressors. In the case where two aberrant females in the Gombe repeatedly seized and ate the babies of other group-members, there was indeed some general distress and alarm, but neither the bereaved mothers nor anyone else attacked them for it nor appeared to change their attitudes to them.[7]

Similarly, individuals who have been quite sharply attacked by others seem usually to feel no resentment. They merely approach the attacker afterwards affectionately, wanting comfort. A reassuring hug from the aggressor seems to restore the social bond, just as it can do with a human toddler of up to about eighteen months.

Indeed, as de Waal most interestingly reports, group-members who have quarrelled invariably do make peace in this way afterwards before night, never letting the sun go down upon their wrath.[8] This strong after-effect makes it all the more interesting that they seem not to respond to offence with that direct indignation which humans begin to feel on their own behalf from about the age of two, and to feel on behalf of those close to them from not much later.

Put in cognitive terms, the difference between the chimp reaction and the human one is a matter of memory. The chimps seem simply to *forget* the offence. But there is surely also an emotional element involved. We humans also forget many things, but even the dimmest of us remember offences – offences both against ourselves and also against others – terribly clearly, in fact we find them atrociously hard to forget. We do not easily get over them; we are passionately interested in them. This – as I am suggesting – is because we are altogether passionately interested in other people's states of mind, in their views of things and especially in their attitudes to ourselves. Even a two-year-old human child can take offence and refuse to be reconciled. This is surely because it does not see an injury done to it just as an injury, but also as a sign of a hostile attitude in the offender.

This intense interest is surely what makes moral generalization possible. A human being does not see a case of infanticide just as an isolated, meaningless event without consequences, as the apes appear to, but as an example of something highly significant. We notice at once the strange attitude that lies behind this act. We are struck by the difference between that attitude and those of other people. We are aware, too, not just of the agent's attitude but also of the attitudes of others involved, such as the bereaved mother. All this, it seems, escapes the chimps, not because they are callous, but because their nervous apparatus is such as to make them only very dimly aware of it.

FREEDOM HAS DEGREES

This means, I think, that there is something misleading about the current tendency to centre the human intellectual distinctiveness which grounds freedom on 'self-consciousness' of some kind. Consciousness of others may be just as necessary. And indeed the two must to some extent develop together. In trying to take decisions, we must become aware both of our own complex, conflicting inner attitudes and also of other people's, and we constantly need to compare the two. Since, however, some non-human animals show elements of both these kinds of awareness, we surely need to accept clearly that freedom is a matter of degree. People are simply much *more* reflective, and therefore more free, than other animals, yet they, too, are by no means wholly reflective or wholly free. As just suggested, the more aware any being is of its internal conflicts, and the more it can, on occasion, distance itself from any of its particular motives, operating as the whole that contains them, the more free it becomes.

If we consider the development of a human baby – something which celibate philosophers have often been unwilling to do – this way of thinking is unavoidable. There is no sudden transition. And the same way of thinking seems appropriate enough when we are trying to understand advanced social animals. As for simpler creatures, the difficulty we have in asking such questions about them seems more a difficulty about being sure what is happening than a flat certainty that they are in no sense free. In such mysterious situations, it is not parsimonious to dogmatize.

HOW CAN INNER CONFLICT DEVELOP?

We need not, I think, pursue the puzzling questions about the exact degree of consciousness found in 'lower animals' here. Current convention deems that they cannot have something rather vaguely called self-awareness, and it is indeed inclined to say that this is a treasure confined to humans. That does not, of course, mean that the simpler creatures lack character altogether. Most animals that have been observed individually at all show distinct characters or dispositions. But when we talk of self-awareness, we do indeed have in mind something more ambitious.

This more ambitious thing clearly does not centre on the kinds of ability that have commonly been tested, such as the power of recognizing spots on one's face in a mirror. It is hard, in fact, to imagine ways in which the more interesting powers that supposedly underlie this one could possibly be tested. Human beings, if they were captured by highly intelligent aliens, would certainly find great difficulty in proving that they possessed these powers.

Perhaps, however, these grander powers may indeed centre on the sort of deliberate effort that we have just been considering – the effort to reinforce or reshape one's central character by forming a constant framework of decision adapted to it, the effort to establish lasting policies with which incoming impulses can agree or conflict. This does call, among other things, for a fairly sophisticated memory and imagination, giving a good sense of the past and future and of alternative pasts and futures. That is a crucial function of our self-knowledge, and no doubt it has to be an important part of the needed increase in intelligence.

Given that clearer intellectual grasp of what is happening within us, how does motivation look? Perhaps it is more like the interaction of currents in the bed of a river than the succession of waves on a beach. Perhaps we move towards the idea that the containing framework is not just an inert river bed, but can to some extent contort itself, changing its shape somewhat and controlling the way the currents flow. Things are now much more complex. More intelligence, more sensibility, was needed to produce this more conscious state, and more still is needed to handle it. In order to deal with its wider range of knowledge and sympathy, the hapless subject needs to keep on becoming yet more thoroughly aware, both of what is happening in it and of what it is trying to do.

This change must (I am suggesting) be a gradual one, not just because Darwin was a gradualist but because cataclysms make no kind of sense here. Morality could not be invented and imposed by a hopeful monster. It cannot be a thunderclap, occurring along with the instant invention of language at the moment of the sudden and final emergence of the human race. The idea that language alone did the trick is a particularly strange one. It surfaces in the way that some optimists seem to expect that teaching apes ASL (i.e. American Sign Language) will enable them to explain their distinctive world-view to us – as if all the human languages that apes might learn were not expressions of the human world-view. The apes are not blank paper on to which language can be stamped. They have a world-view of their own. Without some understanding of it we cannot hope to grasp what they are doing with language.

The distinctively human world-view, however, is so complex that it obviously must have taken many ages to develop, ages during which language has been developing with it. It is not remotely plausible to suggest that there could be two successive moments in the development of a species, one when it has not got language and the next when it has invented the whole of it, and that between these two points it would have made the whole journey from a machine condition to self-awareness and free will. The change has to be much more diffused and more general, a change with many aspects and many roots.

The distance travelled has, then, been vast. Yet that does not mean that we have completely changed our nature. We cannot usefully compare our state with the state of other animals if we still paint that contrast in the lurid, melodramatic black-and-white of Descartes' tradition. The yawning metaphysical gulf between people who are pure subjects and animals that are mere objects was a fantasy. If we want a more realistic approach, we need to listen to those who have paid direct attention to animals, which means using the tradition launched by Darwin. Let us go back now to see what his analysis of moral origins amounts to.

17

MINDS RESIST STREAMLINING

————— •◦• —————

THE IMPERIOUS WORD 'OUGHT'

DARWIN'S PICTURE is striking, and not at all familiar. He shows light being brought for the first time, at the point when the human consciousness of conflict develops, into a huge dark cavern, a place where people who paid little attention to each other have long been carrying on their work separately by the glimmer of a few scattered candles. The new light shows with fearful clarity the relations between enterprises which nobody had ever thought of connecting before.

We may be reminded of Plato's Cave,[1] but the story this time concerns an earlier and deeper predicament. Plato's murky, image-ridden cave represents our whole existing everyday life as a miasma of flickering illusions. But Plato thought we could escape from this cave altogether by following the light of reason. Clear thought could lead us right away from the strife of the desires to a spiritual and intellectual reality beyond it.

Darwin, however, is talking about a much more ancient illumination. He is suggesting how we come to be in our present confused and often odious state in the first place. He is pointing to our natural conflicts of motive as generating a range of dilemmas from which we cannot possibly escape, problems which are still, and always must be, basic to our whole existence. We can partly resolve these problems provided that we understand them, and our efforts to do so lie at the core of our moralities. But there exists no simpler escape route. There is no alternative, pure, conflict-free stream of motivation by which we can bypass these problems entirely. Even the motives that lead us to reason are, after all, themselves a part of our original equipment. Intellectual occupations have by now been much more thoroughly developed and practised than they were in Plato's day, but they have not proved any more immune to misdirection than other ways of life.

At a deep level, these two diagnoses may not be quite so different as they look. Darwin and Plato do agree in being seriously alarmed about human weakness and wickedness, and in attributing these things to deep inner divisions of motive. There is, however, certainly a great difference of direction and emphasis over what should be done about this. Plato expressed the idea of inner division in the Phaedrus myth,[2] where the human soul's chariot is drawn by a bad black horse as well as a good white one, so that the charioteer cannot keep up with the gods and their well-matched teams of white horses. But for Plato, these horses are already named. They are simply black or white, good or bad impulses. The charioteer who hears the Platonic message has a straightforward choice between good and bad, reality and illusion.

On Darwin's view, the need to understand our motives more fully is surely the starting-point. Plato also thought this understanding important, but he believed that it was a relatively simple undertaking. He saw motives as divisible fairly easily into rational and irrational, serious and illusory. Accordingly, a disciplined, ascetic way of life designed to subdue the bad ones under the good ones was the only possible solution. Aristotle, as he gradually developed a more biological approach, abandoned Plato's simple moral dualism. He saw the need for an ideal that involved the whole person, not just a detached rational faculty. He suggested a much more discriminating, less black-and-white way of classifying motives.

Here, as might be expected, Darwin follows him, suggesting that morality necessarily works to harmonize the motives that we have actually got, rather than to impose a quite new pattern. For Darwin, this obscure and alarming workplace, this muddle of conflict-ridden motivation emerging from evolution, is still our home. It is the only mind that we have. It is where we must make our choices and exert our freedom. Though cultures have done their utmost to reorganize it, they have never been able to root out its deep anomalies. Our conflicts are real, not illusory. Our freedom must lie in becoming aware of them and in learning to arbitrate them better.

Though Darwin does make a general suggestion about how the compromises must go, he is not primarily giving directions about which side to take. He is doing something more basic. He is insisting that no creature with inner conflicts of this gravity can avoid taking sides somehow. *This* is what makes morality necessary. There is no semi-paradisal option of simply letting things take their course. Once you realize that you are constantly wrecking your own schemes in the way that the migrating swallow does, you are forced to evolve some sort of priority system and to try to stick to it. That means having a morality. If, therefore, 'immoralism' is taken to mean having no morality, then it is not a possible option.

THE MORAL ATTRACTIONS
OF SIMPLIFICATION

Until lately, this view of Darwin's received remarkably little attention – indeed, people often didn't seem to know that he had written anything at all about ethics. As I have suggested, this was, no doubt, partly because noisier, simpler Social Darwinist views were proclaimed and accepted as a necessary part of a belief in evolution. But the trouble may also have been that, in our tradition, the idea of deep, difficult inner conflict has been closely linked with religious thinking. Enlightenment intellectuals, associating religion with folly, obscurantism, asceticism and political oppression, deeply distrusted the very idea of irremovable inner conflict. They also needed a fairly optimist view about human motivation in order to make the various revolutions that they were proposing look possible.

Throughout the Enlightenment, then, the mainstream of Western thought about motives favoured the opposite assumption – that the human mind is a well-designed monistic device for supplying a single product, probably pleasure, happiness or individual survival. All motives are, then, either disguised wishes for this single end or, if they are really diverse, minor elements within its general domain. In principle, then, it can always arbitrate cleanly between them in case of conflict.

Immense efforts were made to reduce all motivation tidily to this monolithic pattern. The fact that the several main ends suggested, such as pleasure and self-interest, were not actually the *same* end has worried people, but not half enough, because they often overlook it. In Hobbesian egoism and its descendant social-contract theories, the aim is self-preservation. In classical Utilitarianism it is the general happiness or pleasure. And even Freud, who in many ways did lead back towards a more realistic recognition of conflict and complexity, still thought that reason always demanded the reduction of many motives to one wherever possible, and used both hedonism and egoism in determined efforts to achieve this.

It is really not clear why rationality was held to require any such streamlining. Certainly successful reduction of two motives to one can sometimes resolve conflicts. Two possible courses do sometimes turn out to be means to the same end, or parts of the same whole, and this is a happy ending both practically and intellectually. But the process does not always work, and there is no reason why it should be expected to. Hume put great confidence in it, evidently seeing it as part of a liberating Newtonian revolution that would revamp psychology:

> It is entirely agreeable to the rules of philosophy, and even of common reason, where any principle has been found to have a great force and energy in one instance, to ascribe to it a like energy in all similar instances. This indeed is Newton's chief rule in philosophizing.[3]

And again, on this same project of grouping all motives together as parts of the search for utility:

> Thus we have established . . . that it is from natural principles this variety of causes excite pride and humility, and that it is not by a different principle each cause is adapted to its passion. We shall now proceed to enquire how we may *reduce these principles to a lesser number, and find among the causes something common on which their influence depends.*[4] (Emphasis mine)

But there is no *a priori* reason to expect that motives will boil down to a single substrate. The question how many basic motives we have is an empirical one, like the question how many chemical elements there are. In both cases fewer would be neater, but the world is not always neat.

This wish for theoretical neatness converged, then, with the political appeal of motivational monism – with the reaction against the Christian insistence on conflict. Reformers backed worldly motives as a protest against the apathy produced by the ascetic and otherworldly element in Christian thinking. In order to make sure that people fought to get some satisfaction in this world, rather than just despairing and waiting for the next, campaigners called on them to pursue their central worldly interest. Beyond that, there was the usual appeal of a powerful half-truth about psychology, exciting enough to be easily spread beyond its proper limits.

EVOLUTIONARY PUZZLES

This rigid approach, however, made it very hard to see how the theory of evolution, when it arrived, bore on problems of motivation. Since metaphysical disputes were regularly polarized as debates between Religion and Reason, rationalists naturally assumed that evolution would be on their side. Science, they hoped, had now delivered its verdict and revealed the guiding purpose of the human mechanism which they had been seeking. What was it? Herbert Spencer's easy answer caught on at once, and elements of it have been woven into the way we all think about evolution. The very words 'evolution' and 'survival of the fittest' are Spencer's, and their apparent implications about value have given constant trouble.

That simple answer to a complex problem still has enormous appeal for a host of obvious reasons, political, economic and emotional. But perhaps its strongest attraction has been the hope of finality – the prospect of a plain, scientifically certified ruling that could finally unify our nature, showing the painful divisions within it to be just an illusion.

This premature hope blocks the more genuine, serious kind of unification which is actually possible to us. The pretence that we are free from profound clashes does not help us. Acknowledging these clashes is the realistic way to regard the human heritage, the honest answer to the Enlightenment's euphoric project of streamlining our motives. No doubt not all of these conflicts are equally dramatic, but many are very serious. Any creature that develops enough intelligence to become conscious of them does indeed have to think hard about how to resolve them. Indeed, no complete and final solution is possible.

Intelligence, even with all the powers of culture at its disposal, has certainly never enabled our species to clear out its vast cavern, to uproot all the pre-existing emotional structures and start again. And this is probably just as well, since intelligence alone would not have the slightest idea how to generate a whole new set of emotions to replace them.

THE ROLE OF INTELLIGENCE

What, then, does intelligence do? It helps us build a culture – a set of customs expressing a priority system, which will show us how to settle these conflicts in certain agreed ways so as to make the stresses of decision more or less bearable. Within that general framework, it then helps us to decide, by creative effort, how to settle the further clashes that constantly arise and that our culture has not settled. It is because there is no pre-set, universal priority system available that cultures differ so much. Yet it is because their basic problems are still the same that cultures are, none the less, so similar.

Language too, which is so often thought of as the whole secret of human uniqueness, is no doubt a necessary tool in this culture-building, but it is not the whole of it. The psychologist Nick Humphrey has an interesting idea which treats this development very much in the spirit of Darwin. He suggests that the whole flowering of intelligence and language in our species has not been primarily a response to practical needs, not directly a matter of tool-making, but rather a response to the difficulties of handling social relations in communities which were growing more enterprising and co-operative.[5]

This might explain the distressing fact that we go on seeing the problems of our lives still so much in personal terms, and that personal feelings – such as vanity and revenge – still tend to take precedence with us over even the most elementary prudence. Our extreme sociality is a more central feature in us than our abstract intelligence, and this fact needs more attention than it has had in all discussions of our uniqueness. This sociality is by no means an unmixed blessing. It leads us constantly to seek interactions of all kinds, hostile as well as friendly. Even our most benign relations are ambivalent. But it is the temper that we have and we have to make the best of it.

The main point that I think we should accept from Darwin is that we should not expect the psychology of motive to be monistic. Evolutionary considerations come in here to back what realistic observation has always said in opposing over-ambitious theory. We should not go on letting either moralists or speculative scholars streamline the shape of life under the delusion that they are serving reason. For instance, classical Utilitarianism, with its emphasis on the interests of the whole community, grasped an important moral truth, one that was badly needed to correct individualism. But neither of these views has found the all-purpose solution that they claim. Nor is there any need to 'reduce' Utilitarianism to egoism in order to make it respectable. We are beings that naturally care directly for others, as well as for ourselves.

There is no single end for human life. That does not mean that our aims have no sort of relation to one another or that no principles can be found for reconciling them. These aims are not a job lot picked up off the street. They have been developed as parts or aspects of a particular kind of life, namely a human one. Many thought-systems already exist to help us in relating them. Each of us grows up inside at least one such system, and we can always build more. And because many of the basic problems remain the same, these thought-systems have no built-in obsolescence. That is why, when the attempt to look forward merely confuses us, we can often get a sudden light by looking back.

We can in some degree escape our blinkers, finding the different perspectives we need in the thought of other ages, or of other cultures. The idea that we inhabit a definitive 'modern' civilization, an age so finally enlightened that it puts all others out of date, has proved a blind alley. Theorists who have not fully seen what was wrong with it have lately tried to give it a 'post-modern' successor – an outlook that would be equally final, but opposite in being passively sceptical, in denying the very possibility of resolving conflicts. (Not all views that have been called 'post-modernist' take this line, but it has been quite widely seen as central.) But the sweeping exaltation of fashion is equally absurd in both cases. No age has, or could have, a final and universal answer.

CONCLUSION

In this book, I have tried to sketch out a notion of our freedom which will do some kind of justice to the two opposing aspects of it. One aspect is the deep complexity and dividedness of our nature. The other is the equally deep need which each of us feels to act somehow as a unity. When we try, however faintly, to *act* rather than merely letting forces flow through us, we are not just trying to throw off some outside tyranny. Though there may be such a tyranny, the distinctively free effort surely lies in trying to impose unity on the inner conflict, to decide – as a whole person – what to do. That unity is not given. It is a constantly ongoing project, a difficult, essentially incomplete integration which can occupy our whole lives.

All this effort can be looked at from two viewpoints. From one angle – the remote one – such striving is indeed a natural process, simply something that humans characteristically do. It is one more species-specific performance like the migration of birds or the engineering of beavers, a performance whose results can, up to a point, often be predicted. From the other angle – our own – it genuinely is what our experience tells us, namely effort, something active, difficult and internal to the agent involved. There is no illusion about this. From this angle, it is always something dynamic and incomplete, something which can only usefully be talked about in language designed to facilitate it and to help it on towards completion. For this purpose, prediction is often a dangerous irrelevance and – since it can be self-fulfilling – sometimes actively pernicious.

To some extent, theorists have of course always recognized the gap between these viewpoints and also the importance of the subjective angle on inner conflict. But since the full situation is both alarming and confusing, most of them have been unable to resist streamlining it by taking sides far too readily. Moralists, anxious to establish self-control, have usually taken the subjective angle and backed a chosen part of the self to conquer and rule the rest. This is as true of rebellious moralists like Nietzsche as it is of more conventional ones. Indeed, as he often pointed out, rebellion often involves even more drastic self-conquest than conformity. And when freedom became an important ideal, this chosen, rebellious part began to be viewed as the one that was distinctively free.

When, however, the further ideal of a scientific approach took the stage, it began to seem that perhaps things should only be looked at from the remote, objective viewpoint, and moreover only with a view to knowledge. Most confusingly, theoretical, 'scientific' thinking itself became exalted over practical thinking, as if they were alternatives. The search for knowledge began in some confused way to be exalted as supreme morally, as if it were not just one of many excellent activities, but obviously the only one necessary. And from that scientific viewpoint, moral problems are of course no more visible than the behaviour of elephants is to an observer who insists on looking at them through a microscope. This book is an attempt to sketch out, however crudely, a suggestion on how to avoid this confusion between the functions of science and morals.

NOTES

1 INNER DIVISIONS

1 M. Midgley, *Beast and Man*, London, Methuen, 1980.
2 M. Midgley, *Heart and Mind*, London, Methuen, 1983.
3 M. Midgley, *Wickedness*, London, Routledge, 1984.
4 Edward O. Wilson, *On Human Nature*, Cambridge, Mass., Harvard University Press, p. 167.
5 Edward O. Wilson, *Sociobiology, The New Synthesis*, Cambridge, Mass., Harvard University Press 1975, p. 3.
6 Richard Dawkins, *The Selfish Gene*, Oxford, Oxford University Press, 1976, pp. x and 2–3. At this point in the book, no special, technical definition of the word 'selfishness' has been given.
7 See the quotation from Spencer given in Chapter 11, p. 124.
8 T.H. Huxley, *Evolution and Ethics*, (Romanes Lecture) London and New York, Macmillan, 1893, pp. viii, 44, 82.
9 Arthur Peacocke, *God and the New Biology*, London, J.M. Dent and Sons, 1986, p. 89.
10 J. Money et al., 'An examination of some basic sexual concepts: the evidence of human hermaphroditism', Johns Hopkins Hospital, Baltimore, 97, 1955, pp. 301–19. Quoted in Anne Moir and David Jessel, *Brainsex; The Real Difference Between Men and Women*, London, Mandarin, 1991, p. 14, along with an excellent survey of the whole position. For a fuller, rather less provocative discussion with fascinating anthropological backing, see Melvin Konner, *The Tangled Wing: Biological Restraints on the Human Spirit*, London, Heinemann, 1982.
11 For suggestions on less desperate ways of approaching this topic, see my article 'On not being afraid of natural sex differences', in *Feminist Perspectives in Philosophy*, ed. M. Griffiths and M. Whitford, London, Macmillan, 1988, p. 29.
12 N.J.H. Dent, 'Freedom and the Individual', Inaugural Lecture at the University of Birmingham, Birmingham Modern Language Publications, January 1992, p. 7. The advertisement appeared in *The Times Literary Supplement*, 29 Nov. 1991.
13 Ibid., pp. 8 and 5.
14 Ibid., p. 4. I think the barnacles come from Plato's image of the deplorably encrusted sea-god Glaucus in *Republic*, bk x, s. 611.
15 John D. Barrow and Frank R. Tipler, *The Anthropic Cosmological Principle*, Oxford and New York, Oxford University Press, 1986, p. 659. I discuss these proposals further in my *Science as Salvation*, London, Routledge, 1992, chs 2 and 17.
16 Barrow and Tipler, *The Anthropic Cosmological Principle*, p. 658.

2 MISGUIDED DEBATES

1 Thomas Nagel, *The View from Nowhere*, Oxford, Oxford University Press, 1986.
2 David Hume, *Enquiry Concerning the Principles of Morals*, first published 1777, s. 134.
3 This is the theme of my book *Heart and Mind*, London, Methuen, 1981.
4 A matter which I have discussed in *Wisdom, Information and Wonder*, London, Routledge, 1989.
5 Daniel Dennett, *Consciousness Explained*, London, Allen Lane, 1991. Suggestions that Dennett should be prosecuted for his title under the Trades Descriptions Act are attractive, but might call for action over too many other books to be practicable.
6 David Hume, *Treatise on Human Nature*, first published 1739, bk i, pt iv, s. 6.
7 This is a central theme in my *Beast and Man*; see especially pt iv, 'The Marks of Man'.
8 John Locke, *Essay Concerning Human Understanding* bk ii, ch. 21, ss. 14–72.

3 GUIDING VISIONS

1 Ilya Prigogine and Isabelle Stengers, *Order Out of Chaos: Man's New Dialogue with Nature*, London, Fontana, 1985, pp. 7–9.
2 Ibid., pp. xxix-xxx.
3 Lewis Wolpert, *The Unnatural Nature of Science*, London, Faber & Faber, 1992, pp. 125, 135, 138.
4 Prigogine and Stengers, *Order Out of Chaos*, p. 3.
5 This terminus of thought was notoriously well expressed by Jacques Monod in *Chance and Necessity*, Glasgow, Collins, Fontana, 1974, especially in his last chapter.
6 Bishop George Berkeley, *The Principles of Human Knowledge*, first published in 1710, s. 51.
7 Raymond Tallis, *The Explicit Animal*, Basingstoke and London, Macmillan, 1991, pp. 251–2.
8 B.F. Skinner, *Beyond Freedom and Dignity*, Harmondsworth, Penguin, 1973, p. 10.
9 Werner Heisenberg, *Physics and Philosophy*, London, Penguin, 1989, pp. 188–9.
10 Rom Harré, David Clarke and Nicola de Carlo, *Motives and Mechanisms*, London and New York, Methuen, 1985, pp. 14 and 16.
11 Ludwig Wittgenstein, *Philosophical Investigations*, trans. G.E.M. Anscombe, Oxford, Basil Blackwell, 1963, p. 8, s. 18.
12 Harré et al., *Motives and Mechanisms*, p. 16.
13 This point is very well discussed by Arthur Peacocke in *God and the New Biology*, London, J.M. Dent & Sons, 1986, p. 6.
14 See Arthur Peacocke, *God and the New Biology*, ch. 2 'Is biology nothing but physics and chemistry?' pp. 21–31.
15 Francis Crick, *What Mad Pursuit?*, London, Penguin, 1989, pp. 138–9.
16 J. Bentham, *Principles of Morals and Legislation*, first published 1789, ch. 1, ss. 1 and 10.
17 F. Nietzsche, *The Will to Power*, para. 1067, and *Beyond Good and Evil*, para. 269; cf. para. 36.
18 I. Kant, *Critique of Aesthetic Judgment*, trans. James Meredith, Oxford, Clarendon Press, 1911, p. 176, 'The faculties of the mind which constitute genius'.
19 A point well discussed by Aristotle, *Nicomachean Ethics*, bk x, ch. 7.

4 HOPES OF SIMPLICITY

1 See ch. 2, pp. 13–15.
2 Francis Crick, *Of Molecules and Men*, Seattle, University of Washington Press, 1966, p. 10.
3 Lewis Wolpert, *The Unnatural Nature of Science*, London, Faber & Faber, 1992, p. 135.
4 Philip R. Slavney and Paul R. McHugh, *Psychiatric Polarities*, Baltimore and London, Johns Hopkins University Press, 1987, p. 8.
5 Ibid., p. 123.

5 CRUSADES, LEGITIMATE AND OTHERWISE

1 Sigmund Freud, paper on 'Narcissism', *Collected Works*, translated under the editorship of James Strachey, Hogarth Press 1953-74, vol. XIV, p. 91.
2 Lewis Wolpert, *The Unnatural Nature of Science*, London, Faber & Faber, 1992, p. 148.
3 Rom Harré, David Clarke and Nicola de Carlo, *Motives and Mechanisms*, London and New York, Methuen, 1985, p. 100.
4 Thomas Hobbes, *Leviathan*, first published 1651, pt 1, ch. 6.
5 E. McBride, 'Social biology and birth control', *Nature*, vol. 113 (31 May 1924), p. 774. Quoted with an excellent discussion by Gary Werskey in *The Visible College*, London, Free Association Books, 1988, p. 32. See also his index, s.v. Eugenics.
6 E. McBride, 'Birth control and human biology', *Nature*, vol. 127 (4 April 1931), p. 511.
7 E. McBride, 'Cultivation of the unfit', *Nature*, vol. 137 (11 January 1936), p. 45.

6 CONVERGENT EXPLANATIONS AND THEIR USES

1 See the opening chapters of Aristotle's *Metaphysics*.
2 Thomas Nagel, *The View from Nowhere*, New York and Oxford, Oxford University Press, 1986, p. 3.
3 Ibid., p. 4.
4 Daniel C. Dennett, *The Intentional Stance*, Cambridge, Mass., Bradford Books, 1987, p. 5.

7 TROUBLES OF THE LINEAR PATTERN

1 Edward O.Wilson, *On Human Nature*, London and Cambridge, Mass., Harvard University Press, 1978, p. 7.
2 See ch. 3, p. 37.
3 Wilson, *On Human Nature*, pp. 10 and 204.
4 Edward O. Wilson, *Sociobiology: the New Synthesis*, Cambridge, Mass., Harvard Belknap Press, 1975, p. 562.
5 Ibid., pp. 6–7.
6 Ibid., p. 4.
7 Ibid., p. 575.
8 See for instance my *Heart and Mind*, ch. 1, 'The human heart and other organs', pp. 15–24.

9   Francis Crick, *What Mad Pursuit? A Personal View of Scientific Discovery*, London, Penguin, 1990.
10 See Wilson, *On Human Nature*, index, s.v. Religion.
11 *New Statesman*, 18 December 1992, pp. 42–5.

8   FATALISM AND PREDICTABILITY

1   Peter Strawson, 'Freedom and resentment', in his book of the same title, London, Methuen, 1974, p. 1.
2   Ibid., p. 23.
3   Ibid., p. 25.
4   Bernard Williams, 'Moral luck' in the book of the same name, Cambridge, Cambridge University Press, 1981, pp. 20–40. See also a recent article on 'Nietzsche's minimalist moral psychology' in the *European Journal of Philosophy*, vol. 1, no. 1 (April 1993), endorsing essentially the same position.
5   Williams, *Moral Luck*, p. 27.
6   See Mary Midgley, *Wickedness*, London, Routledge, 1984, ch. 2, 'Intelligibility and immoralism'.
7   In an article called 'The flight from blame' in *Philosophy*, vol. 62, no. 241, (July 1987), pp. 271–91, part of which is reworked in *Wickedness*, ch. 3, 'The elusiveness of responsibility'.
8   See Choderlos de Laclos, *Les Liaisons Dangereuses*, Letters CXLI and CXLII.
9   Ibid., Letter CXLV.
10 Eric Berne, *Games People Play: The Psychology of Human Relationships*, Harmondsworth, Penguin, 1964, pp. 76–80.
11 Edward O. Wilson, *Sociobiology: The Modern Synthesis*, Harvard University Press, 1975, p. 3.
12 Richard Dawkins, *The Selfish Gene*, Oxford, Oxford University Press, 1976, p. x.
13 A point fascinatingly discussed by Gillian Beer in *Darwin's Plots*, London, Routledge, 1983 and Robert Young, *Darwin's Metaphor*, Cambridge, Cambridge University Press, 1985.
14 Edward O. Wilson, *On Human Nature*, Cambridge, Mass., Harvard University Press, 1978, pp. 2, 3–4.
15 Ibid., p. 167.

9   AGENCY AND ETHICS

1   Cited by Raymond Tallis in *The Explicit Animal*, London, Macmillan, 1991, p. 244, giving reference to 'Emanational Physics', an unpublished manuscript by Stephen Harrison. Tallis says no further details are given. The passage seems, however, to express Bohm's habitual views and does not appear in any way suspect.
2   Rom Harré, David Clarke and Nicola de Carlo, *Motives and Mechanisms*, London and New York, Methuen, 1985, p. 9.
3   B.F. Skinner, *Beyond Freedom and Dignity*, Harmondsworth, Penguin, 1973, pp. 25 and 206.
4   Ibid., p. 20.
5   Good sources for these views are P.N. Johnson-Laird, *The Computer and the Mind*, London, Fontana, 1988; Colin Blakemore and Susan Greenfield (eds), *Mindwaves*, Oxford, Basil Blackwell, 1987, and Paul Churchland, *Matter and Consciousness*, rev. edn, Cambridge, Mass., MIT Press, 1987.

6 Churchland, *Matter and Consciousness*, p. 5.
7 Tallis, *The Explicit Animal*, p. 82.
8 Ibid., p. 83.
9 See ch. 10, pp. 111–15.
10 Skinner, *Beyond Freedom and Dignity*, pp. 101–2.
11 Ibid., p. 114.
12 He has explained how he thinks this should be done in *Causal Powers* by R. Harré and E.H. Madden, Oxford, Basil Blackwell, 1975.
13 Harré et al., *Motives and Mechanisms*, p. 10.
14 Carl Sagan and Ann Druyan, *Shadows of Forgotten Ancestors: A Search for Who We Are*, London, Century, 1991, opening passage.

10 MODERN MYTHS

1 J.J. Rousseau 'Discourse on the origin of inequality' in *The Social Contract and Discourses*, trans. G.D.H. Cole, Everyman's Library, London, Dent & Dutton, 1935, pp. 188, 194, 200.
2 Thomas Hobbes, *Leviathan*, first published 1651, pt 1, ch. 13, Everyman's Library, London, Dent & Dutton, 1931, p. 64.
3 See his *Social Contract*, bk 1, ch. 4.
4 J.-P. Sartre, *Existentialism and Humanism*, trans. Philip Mairet, London, Methuen, 1948, pp. 28, 34, 45, 54.
5 Tim Radford, 'Made to measure', *Guardian*, 3 Jan. 1993, p. 16.
6 T.H. Huxley, 'Evidence as to Man's place in Nature', in *Select Works of Thomas H. Huxley*, New York, John B. Alden, 1886, p. 234.

11 THE STRENGTH OF INDIVIDUALISM

1 J. Thibaut and H. Kelley, *The Social Psychology of Groups*, New York and Chichester, Wiley, 1959, p. 37.
2 Thomas Nagel, *The View from Nowhere*, Oxford and New York, Oxford University Press, 1986.
3 Plato, *Republic*, bk 1, 352d.
4 For its bad consequences today see Robert Bellah et al., *Habits of the Heart*, London, Hutchinson, 1985, and, on the absurdity of exalting competition over co-operation on the human scene, Alfie Kohn, *No Contest: The Case Against Competition*, Boston, Houghton Mifflin, 1986.
5 Herbert Spencer, *Social Statics*, New York, D. Appleton & Co., 1864, pp. 414–15. Quoted in Richard Hofstadter, *Social Darwinism in American Thought*, New York, George Braziller, 1959.
6 H. Spencer, *Social Statics*, pp. 79–80. Quoted in *Social Darwinism in American Thought*, p. 40.
7 H. Spencer, *Data of Ethics*, preface. Quoted in Hofstadter, *Social Darwinism in American Thought*, p. 40.
8 Ilya Prigogine and Isabelle Stengers, *Order Out of Chaos*, London, Fontana, 1985, pp. 12–14.

12 THE RETREAT FROM THE NATURAL WORLD

1 See the opening chapters of my *Beast and Man*, London, Methuen, 1980.
2 I. Kant, 'Duties towards animals and spirits', in *Lectures on Ethics*, trans. Louis Infield, London, Methuen, 1930.
3 See Sy Montgomery, *Walking with the Great Apes*, Boston, Mass., Houghton

Mifflin, 1991.

4 Peter Atkins, *The Creation*, Oxford and San Francisco, W.H. Freeman, 1987, p. 53.

5 *Leviathan* pt 1, ch. 15.

6 S. Freud 'Narcissism', in *Collected Works*, translated under the editorship of James Strachey, Hogarth Press 1953–74, vol. XIV, p. 91.

7 David Barash, *Sociobiology: The Whisperings Within*, London, Souvenir Press, 1980, p. 3.

8 *The Tangled Wing*, London, Heinemann, 1982.

13 HOW FAR DOES SOCIABILITY TAKE US?

1 Charles Darwin, *The Descent of Man*, first published 1871, reprinted Princeton, Princeton University Press, 1981, vol. 1, pt 1, ch. 3.

2 See Robert Hinde, 'Ethological models and the concept of drive', *British Journal for the Philosophy of Science*, 6 (1956), p. 321.

3 On the very complex primate situation, see A. Desmond, *The Ape's Reflexion*, London, Blond & Briggs, 1979.

4 Darwin, *The Descent of Man*, pp. 84, 91.

5 Ibid., p. 72.

6 Ibid., p. 92.

7 Ibid., p. 106.

14 THE USES OF SYMPATHY

1 Georgia Mason, 'Role-swapping makes monkeys of macaques' in *New Scientist*, vol. 137, no. 1858 (30 January 1993), summarizing an article by Daniel Povinelli in *Animal Behaviour*, vol. 44, p. 269.

2 J.-J. Rousseau, 'Discourse on the origin of inequality', in *The Social Contract and Discourses*, trans. G.D.H. Cole, Everyman's Library, London, Dent & Dutton, 1935, p. 200.

3 David Hume, *Enquiry Concerning the Principles of Morals*, first published 1751, ss. 174–250.

4 For instance in *Beyond Good and Evil*, s. 32.

5 See Melvin Konner, *The Tangled Wing: Biological Constraints on the Human Spirit*, London, Heinemann, 1982; Margaret Mead, *New Lives for Old*, London, Gollancz, 1956.

6 For Callicles, see Plato's *Gorgias*. ss. 481–527. For Thrasymachus, see his *Republic*, bk 1, ss. 336–54.

7 Frans de Waal, *Chimpanzee Politics*, Baltimore and London, Johns Hopkins University Press, 1989, p. 207.

8 Bks IV–V, ss. 445–71.

9 See Bernard Williams's paper, 'Moral luck', in his book of the same title, Cambridge, Cambridge University Press, 1981, p. 20.

10 See ch. 8, pp. 85–7.

11 Bishop Joseph Butler, 'Dissertation upon the nature of virtue', printed with his *Fifteen Sermons*, London, G. Bell & Sons, 1959, p. 247.

12 I discussed it more fully in pt 1 of my *Beast and Man*, London, Methuen, 1980.

13 P. Foot, 'When is a principle a moral principle?', *Proceedings of the Aristotelian Society*, supplementary volume, 1954.

15 ON BEING TERRESTRIAL

1 I. Kant, *The Moral Law*, trans. H.J. Paton, London, Hutchinson University Library, 1948, ch. 1, s. 5, p. 61.
2 See ch. 2, pp. 21–3.
3 Werner Heisenberg, *Physics and Philosophy*, Harmondsworth, Penguin, 1989, pp. 189–90.

16 WHAT KIND OF BEINGS ARE FREE?

1 David Hume, *Enquiry Concerning Human Understanding*, first published 1748, ch. 8.
2 Konrad Lorenz, *On Aggression* (trans. Marjorie Latzske), London, Methuen, 1966, pp. 96–7.
3 These issues have been well discussed by Donald Griffin in *The Question of Animal Awareness*, New York, Rockefeller University Press, 1976, and *Animal Thought*, Cambridge, Mass., Harvard University Press, 1984.
4 Adrian Desmond, *The Ape's Reflexion*, London, Blond & Briggs, 1979, p. 220
5 Frans de Waal, *Chimpanzee Politics*, Baltimore and London, Johns Hopkins University Press, 1989, p. 206.
6 Ibid., pp. 124–6.
7 Desmond, *The Ape's Reflexion*, pp. 221–2 and 244. See also Jane Goodall, *Through a Window*, London, Weidenfeld & Nicolson, 1990, pp. 64–6.
8 de Waal, *Chimpanzee Politics*, p. 41.

17 MINDS RESIST STREAMLINING

1 *Republic*, bk 7, ss. 514–21.
2 Plato, *Phaedrus*, ss. 246–57.
3 David Hume, *Enquiry Concerning the Principles of Morals*, first published 1751, s. 163.
4 David Hume, *Treatise of Human Nature*, first published 1739, bk 2, pt 1, s. 4.
5 Nicholas Humphrey, 'Nature's psychologists', *New Scientist*, 29 June 1978, and an expanded version in *Consciousness and the Physical World*, ed. B.D. Josephson and V.S. Ramachandran, London, Pergamon Press, 1979.

INDEX

———— •◆• ————

Aldebaran 138, 149
American Sign Language 175
Antaeus 118
Aristotle 10, 136, 165, 178
Arjuna 107
Atkins, Peter 132

Barrow, John 10, 12, 95
Bentham, Jeremy 38–40, 84
Berkeley, Bishop George 33
Bernal, J.D. 61
Berne, Eric 86
Blake, William 145, 158
Bohm, David 96
Brave New World 169
Butler, Bishop Joseph 152

Callicles 145
Calvin, John 86
Chicago 5
Christianity 6, 77, 110, 114, 130
Churchland, Paul 100
Copernicus, N. 33
Crick, Francis 38, 45, 78, 130, 156

Daleks 138
Darwin, Charles 5–6, 20, 75, 87–8,
    115–16, 120, 139–45, 163, 172–82
Darwinism 3, 5, 17, 136, 139
Dawkins, Richard 5, 78, 86, 167
de Waal, Frans 131, 146, 172–3
Democritus 30
Dennett, D. 21, 67
Dent, Nicholas 8–9
Descartes, René 7, 11, 20, 28, 32, 74,
    102, 167, 176
Desmond, Adrian 171–2
DNA 5, 86, 89–90

Einstein, Albert 88
Epicurus 30
Eros 16
Eugenics Society 61

Foot, Philippa 153
Fossey, Dian 131
Freud, Sigmund 16, 21, 45,
    52, 59, 76, 79–80, 133, 136, 179

Galdikas, Biruté 131
Galileo 34, 55, 58
Genesis, Book of 110, 119
God 30, 32, 34, 48–9, 78, 86, 97, 100,
    110–11, 169
Gombe 171–2
Goodall, Jane 131
Grand Canyon 101
Guardian 112

Hamlet 107
Harré, Rom 34–6, 57, 96, 105
Hegel, G.W.F. 9, 66
Heisenberg, Werner 34, 168
Heraclitus 16, 126
Hobbes, Thomas 5, 17, 48, 52, 58,
    78, 109–10, 121–2, 125, 132–3,
    138, 149, 166, 179
Hoyle, F. 130
Hume, David 14, 21–2, 66, 77, 136,
    143–5, 166, 170, 180
Humphrey (chimpanzee) 171
Humphrey, Nicolas 182
Huxley, Julian 131
Huxley, Thomas Henry 6–7, 16,
    115–16, 129
Hydra 19

Jones, Evan 45, 63–71, 76, 86
Jung, C.G. 145

Kant, Immanuel 41, 111, 114, 131, 136, 145–6, 148–52, 157, 166
Kelley, H. 121
Konner, Melvin 134

Lenoble, R. 30
Liaisons Dangereuses, Les 85
Locke, John 24
Lorenz, Konrad 131
Lucretius 30

Machiavelli, N. 145
Manichees 16
Marx, Karl 40, 45, 61, 76–7, 125
Mendel, Gregor 75
Mike (chimpanzee) 171
Mill, J.S. 158
Money, John 8
Moore, G.E. 18
Morris, Desmond 3, 128, 131

Nagel, Thomas 13, 66, 121
Nature 61–1
Nazism 61
Newton, Sir Isaac 15, 28, 34, 36, 55, 96, 166, 180
Nietzsche, F. 20, 39–40, 45–6, 52–3, 79, 107, 122–4, 144, 166, 184

Occam's Razor 38
Orestes 107

Peacocke, Arthur 7, 38
Persephone 117
Phaedrus 1378
Plato 14–15, 18, 121, 130, 145, 148
Prigogine, Ilya 27–8, 30, 127
Procrustes 14, 56, 69, 152

Rawls, John 112
Roget's Thesaurus 102
Rousseau J.J. 109, 119, 143
Russell, Bertrand 18
Russia 49

Sartre, J.P. 8, 11, 20, 102, 111–15, 124, 150, 166
Skinner, B.F. 19, 34, 36, 45, 48, 73–6, 84, 97–9, 104
Social Darwinists 5–6, 88, 124–6, 129, 179
Socrates 157
Spencer, Herbert 6, 19, 124–5, 129, 181
Spinoza, B. 157
Stengers, Isabelle 27–8, 30, 127
Strawson, Peter 82–3
Strum, Shirley 131

Tallis, Raymond 33, 101
Thanatos 16
Theaetetus 18
Thibaut, J. 121
Thrasymachus 145
Tinbergen, Niko 131
Tipler, Frank R. 10, 12, 95
Turing, Alan 10

USA 57, 125

Voltaire 77

Wallace, Alfred 7
Watson, J.B. 48
Werskey, Gary 61
Williams, Bernard 83–4, 149
Wilson, E.O. 71–5, 78, 99
Wittgenstein, L. 35
Wolpert, Lewis 29–30, 48–9, 54, 57

Printed in the United States
by Baker & Taylor Publisher Services